Student Solutions Manual
for Kaseberg's

INTRODUCTORY
ALGEBRA

A JUST-IN-TIME APPROACH

Cindy Rubash

Brooks/Cole
Thomson Learning.

Australia • Canada • Denmark • Japan • Mexico • New Zealand • Philippines
Puerto Rico • Singapore • Spain • United Kingdom • United States

Project Development Editor: *Michelle Paolucci*
Senior Editorial Assistant: *Erin Wickersham*
Marketing Team: *Leah Thomson, Debra Johnston*
Production Coordinator: *Dorothy Bell*

Cover Design: *Elise S. Kaiser*
Cover Illustration: *Don Baker*
Print Buyer: *Micky Lawler*
Printing and Binding: *Globus Printing*

For more information, contact:
BROOKS/COLE
511 Forest Lodge Road
Pacific Grove, CA 93950 USA
www.brookscole.com

For permission to use material from this work, contact us by
web: www.thomsonrights.com
fax: 1-800-730-2215
phone: 1-800-730-2214

Printed in the United States of America

10 9 8 7 6 5 4 3

ISBN 0-534-37325-9

CONTENTS

Section 1.1

1. For example: Assume you work 20 days a month or you work every day of the month.

3. For example: Assume your teacher knows your last name and in which class you are enrolled.

5. 3 panels for 1st pen, 2 additional panels for each pen after that.

$$3 + 2 \cdot 19 = 41$$

7. 3 panels for first 2, 2 panels for every 2 displays thereafter.

$$3 + 2 \cdot 9 = 21$$

9. One table will seat 5, 2 tables (as shown) will seat 8, each additional table adds 3 seats.

5	(1 table)
$5 + 1 \cdot 3 = 8$	(2 tables)
$5 + 2 \cdot 3 = 11$	(3 tables)
$5 + 3 \cdot 3 = 14$	(4 tables)

15. The surgeon is the boy's mother.

Section 1.2

1. $4 \cdot \dfrac{1}{2} = \dfrac{4}{2} = 2$

3. $\dfrac{1}{3} + \dfrac{1}{2} = \dfrac{2}{6} + \dfrac{3}{6} = \dfrac{5}{6} \approx 0.83$

5. $5 \div 15 = \dfrac{5}{15} = \dfrac{1}{3} \approx 0.33$

7. $\dfrac{3}{5} \cdot \dfrac{4}{9} = \dfrac{3 \cdot 4}{5 \cdot 9} = \dfrac{12}{45} = \dfrac{4}{15} \approx 0.27$

9. $2\dfrac{2}{3} - 1\dfrac{1}{4} = \dfrac{8}{3} - \dfrac{5}{4} = \dfrac{32}{12} - \dfrac{15}{12} =$

$\dfrac{17}{12} = 1\dfrac{5}{12} \approx 1.42$

11. $2\dfrac{2}{3} \div 1\dfrac{1}{4} = \dfrac{8}{3} \div \dfrac{5}{4} = \dfrac{8}{3} \cdot \dfrac{4}{5} = \dfrac{32}{15} =$

$2\dfrac{2}{15} \approx 2.13$

13. $\dfrac{4}{5} - \dfrac{1}{3} = \dfrac{12}{15} - \dfrac{5}{15} = \dfrac{7}{15} \approx 0.47$

15. $1\dfrac{1}{2} + 2\dfrac{1}{4} = \dfrac{3}{2} + \dfrac{9}{4} = \dfrac{6}{4} + \dfrac{9}{4} =$

$\dfrac{15}{4} = 3\dfrac{3}{4} = 3.75$

17. Division by 10 is the same as

multiplication by $\dfrac{1}{10}$

$\dfrac{1}{10}(1.5 + 2.5) = \dfrac{1}{10}(4.0) = 0.4$

$0.4 = \dfrac{4}{10} = \dfrac{2}{5}$

19. $\dfrac{1}{10}(20 - 15) = \dfrac{1}{10}(5) = 0.5$

$0.5 = \dfrac{5}{10} = \dfrac{1}{2}$

21. $12 \cdot \$48 = \$576$

23. $\$404 - 20 = \$384$ (tuition cost)

$\$384 \div \dfrac{\$48}{\text{cr. hr.}} = 8 \text{ credit hours}$

25. a. small $= 1\dfrac{1}{2}$ yd, large $= 1\dfrac{3}{4}$ yd

$1\dfrac{1}{2} + 1\dfrac{3}{4} = \dfrac{3}{2} + \dfrac{7}{4} = \dfrac{6}{4} + \dfrac{7}{4} = 3\dfrac{1}{4}$ yd

b. $5 \cdot \left(1\dfrac{3}{4}\right) = 5 \cdot \dfrac{7}{4} = \dfrac{35}{3} = 8\dfrac{3}{4}$ yd

27. a. ex. large $= 2\dfrac{1}{8}$ yd, large $= 1\dfrac{3}{4}$ yd

$2\dfrac{1}{8} - 1\dfrac{3}{4} = \dfrac{17}{8} - \dfrac{7}{4} = \dfrac{17}{8} - \dfrac{14}{8} = \dfrac{3}{8}$ yd

27. b. $21 \text{ yd} \div 1\dfrac{3}{4} \text{ yd} = 21 \div \dfrac{7}{4} = 21 \cdot \dfrac{4}{7} =$

$\dfrac{84}{7} = 12 \text{ shirts}$

Section 1.2 (con't)

29.

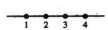

Input (dots)	Output (line seg.)
1	2
2	3
3	4
4	5

Output is one more than the input.

Output is 11 when input is 10.

31.

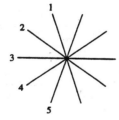

Input (lines)	Output (segments)
1	2
2	4
3	6
4	8
5	10

Output is twice the input. Output will be 20 when input is 10.

39.

41.

43. Payment is 10% of $375.

$$\frac{1}{10} \cdot \$375 = \$37.50$$

45. The payment is over $50 so the balance is over $500;

$75 - $50 = $25 (excess over $500)

Balance is $500 + $25 = $525

47.

Input	Output
0	5
1	2(1) = 2
2	5
3	2(3) = 6
4	5
5	2(5) = 10
6	5
7	2(7) = 14
8	5

49.

Input cr. hr.	Output cost
8	8($55) = $440
9	9($55) = $495
10	10($55) = $550
11	$600
12	$600

Section 1.2 (con't)

53. a. Natural numbers are integers that are not negative or zero.

b. Negative real numbers can be used to write elevations below sea level.

53. c. Rational numbers.

d. Integers.

55. a. $3+9=12$, $3\cdot 9 = 27$

b. $3+7=10$, $3\cdot 7 = 21$

57.

a	b	$a+b$	$a-b$
15	5	$15+5=20$	$15-5=10$
$\frac{3}{4}$	$\frac{2}{5}$	$\frac{3}{4}+\frac{2}{5}=\frac{15+8}{20}=1\frac{3}{20}$	$\frac{3}{4}-\frac{2}{5}=\frac{15-8}{20}=\frac{7}{20}$
0.36	0.06	$0.36+0.06=0.42$	$0.36-0.06=0.30$
5.6	0.7	$5.6+0.7=6.3$	$5.6-0.7=4.9$
2.25	1.5	$2.25+1.5=3.75$	$2.25-1.5=0.75$

a	b	$a\cdot b$	$a\div b$
15	5	$15\cdot 5=75$	$15\div 5=3$
$\frac{3}{4}$	$\frac{2}{5}$	$\frac{3}{4}\cdot\frac{2}{5}=\frac{6}{20}=\frac{3}{10}$	$\frac{3}{4}\div\frac{2}{5}=\frac{3}{4}\cdot\frac{5}{2}=\frac{15}{8}=1\frac{7}{8}$
0.36	0.06	$0.36\cdot 0.06=0.0216$	$0.36\div 0.06=6$
5.6	0.7	$5.6\cdot 0.7=3.92$	$5.6\div 0.7=8$
2.25	1.5	$2.25\cdot 1.5=3.375$	$2.25\div 1.5=1.5$

59. a. $\dfrac{15\cdot 100}{100}=\dfrac{1500}{100}=1500\%$

$\dfrac{3}{4}=0.75,\ \dfrac{0.75\cdot 100}{100}=\dfrac{75}{100}=75\%$

$\dfrac{0.36\cdot 100}{100}=\dfrac{36}{100}=36\%$

$\dfrac{5.6\cdot 100}{100}=\dfrac{560}{100}=560\%$

$\dfrac{2.25\cdot 100}{100}=\dfrac{225}{100}=225\%$

59. b. $\dfrac{5\cdot 100}{100}=\dfrac{500}{100}=500\%$

$\dfrac{2}{5}=0.4,\ \dfrac{0.4\cdot 100}{100}=\dfrac{40}{100}=40\%$

$\dfrac{0.06\cdot 100}{100}=\dfrac{6}{100}=6\%$

$\dfrac{0.7\cdot 100}{100}=\dfrac{70}{100}=70$

$\dfrac{1.5\cdot 100}{100}=\dfrac{150}{100}=150\%$

Mid-Chapter 1 Test

1.

The number of triangles is 2 less than the number of sides.

Input sides	Output triangles
3	1
4	2
5	3
6	4
7	5
20	20 - 2 = 18

2. a. True

b. True

c. False, zero is added to the set of natural numbers to make the set of whole numbers.

d. False, 3 divided by 4 is not an integer.

3. a. $2\dfrac{1}{3}+3\dfrac{3}{4}=\dfrac{7}{3}+\dfrac{15}{4}=\dfrac{28}{12}+\dfrac{45}{12}=$

$\dfrac{73}{12}=6\dfrac{1}{12}$

b. $1\dfrac{1}{2}\cdot 1\dfrac{1}{4}=\dfrac{3}{2}\cdot\dfrac{5}{4}=\dfrac{15}{8}=1\dfrac{7}{8}$

c. $3\dfrac{1}{4}-1\dfrac{3}{4}=\dfrac{13}{4}-\dfrac{7}{4}=\dfrac{6}{4}=1\dfrac{1}{2}$

d. $3\dfrac{1}{2}\div 1\dfrac{3}{4}=\dfrac{7}{2}\div\dfrac{7}{4}=\dfrac{7}{2}\cdot\dfrac{4}{7}=2$

4. a. $\$1.50+2\cdot\$1=\$1.50+\$2=\$3.50$

b. $\$1.35+2\cdot\$0.90=$

$\$1.35+\$1.80=\$3.15$

c. $\$1.05+4\cdot\$0.70=$

$\$1.05+\$2.80=\$3.85$

d. Cost of first minute:

$\$5.25 - \$1.05 = \$4.20$

Remaining minutes @ \$0.70 per minute: $\$4.20 \div \$0.70 = 6$

$1 + 6 = 7$ minutes

Section 1.3

1. a. $3 \cdot n = 3n$

b. $8 \div n = \dfrac{8}{n}$

c. $n - 4$

d. $n \div 5 = \dfrac{n}{5}$

e. $15 \cdot \$n = \$15n$

3. a. $3 + 2 \cdot n = 3 + 2n$

b. $4 - 3 \cdot n = 4 - 3n$

c. $7 \cdot n + 4 = 7n + 4$

d. $n \cdot n = n^2$

e. $\$0.79 \cdot n = \$0.79n$

5. a. $2\pi r$; 2 and π are both constants and numerical coefficients, r is a variable.

b. 1.5x; 1.5 is a constant and a numerical coefficient, x is a variable.

c. -4n + 3; -4 is a constant and a numerical coefficient, 3 is a constant, n is a variable.

d. x^2 - 9; The implied 1 (before the x^2) is a constant and numerical coefficient, -9 is a constant, x is a variable.

11. a. 35% of $n = \dfrac{35}{100} \cdot n = 0.35n$

b. 10% of $x = \dfrac{10}{100} \cdot x = 0.10x$

c. $87\frac{1}{2}$% of $n = \dfrac{87.5}{100} \cdot n = 0.875n$

d. $37\frac{1}{2}$% of $x = \dfrac{37.5}{100} \cdot x = 0.375x$

e. $\frac{1}{2}$% of $n = \dfrac{0.5}{100} \cdot n = 0.005n$

f. 108% of $x = \dfrac{108}{100} \cdot x = 1.08x$

13. Let x = input; output = $3x + 2$

Input	Output
1	$3(1) + 2 = 5$
2	$3(2) + 2 = 8$
3	$3(3) + 2 = 11$
4	$3(4) + 2 = 14$
5	$3(5) + 2 = 17$
x	$3x + 2$

Section 1.3 (con't)

15.

Input	Output
4	$2(4) + 1 = 9$
50	$2(50) + 1 = 101$
100	$2(100) + 1 = 201$
n	$2n + 1$

figure c; twice the input plus 1

17.

Input	Output
4	$4(4) - 1 = 15$
50	$4(50) - 1 = 199$
100	$4(100) - 1 = 399$
n	$4n-1$

figure b; one less than 4 times the input.

19.

Input t hrs	Output D miles
1	$55(1)=55$
2	$55(2)=110$
3	$55(3)=165$
t	$55t$

21.

Input n sales	Output income \$
1	$250+75(1)=325$
2	$250+75(2)=400$
3	$250+75(3)=475$
n	$250+75n$

23. Not necessarily, x is a variable and may represent any positive or negative number or zero.

25.

Input	Output
-6	0
-5	2
-4	2
-3	0
-2	2
-1	2
0	0
1	2
2	2
3	0
4	2
5	2
6	0

Section 1.3 (con't)

27.

Input	Output
-3	0
-2	0
-1	0
0	0
1	1
2	2
3	3

29.

Input	Output
8	9
50	51
101	202
Even n	$n + 1$
Odd n	$2n$

31.

Pairs of Pens	Panels
4	$5(4) + 2 = 22$
x	$5x + 2$
10	$5(10) + 2 = 52$
25	$5(25) + 2 = 127$

Rule: 2 more than 5 times the input.

33.

Tables	Chairs
2	8
4	$3(4) + 2 = 14$
n	$3n + 2$
10	$3(10) + 2 = 32$

35.

x	y	xy	$x + y$
4	3	12	$4 + 3 = 7$
1	12	$1 \cdot 12 = 12$	$1 + 12 = 13$
5	4	$5 \cdot 4 = 20$	$5 + 4 = 9$
20	1	20	$1 + 20 = 21$
2	2	$2 \cdot 2 = 4$	$2 + 2 = 4$
6	3	$6 \cdot 3 = 18$	$6 + 3 = 9$

Section 1.4

1. A (-2, 2)

B (-5, 0)

C (-3, -4)

D (0, -2)

E (4, -4)

F (2, -6)

G (4, 0)

H (4, 6)

I (0, 5)

3. a. Quadrant 2 (x is negative, y is positive)

b. Quadrant 4 (x is positive, y is negative)

c. Quadrant 3 (both x & y are negative)

d. Quadrant 3 (both x & y are negative)

5. a. vertical-axis ($x = 0$)

b. vertical-axis ($x = 0$)

c. horizontal-axis ($y = 0$)

7.

Input	Output
x	y
0	0
-2	1
-4	2
2	-1
4	-2
-6	3
5	-2.5

9.

Input	Output
x	$y = 2x + 5$
0	$2(0) + 5 = 5$
1	$2(1) + 5 = 7$
2	$2(2) + 5 = 9$
3	$2(3) + 5 = 11$
4	$2(4) + 5 = 13$
5	$2(5) + 5 = 15$

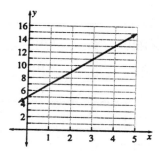

11.

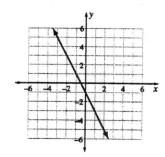

13.

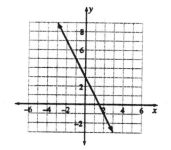

Section 1.4 (con't)

15 - 21

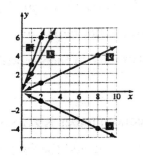

23. a.

Input x lbs.	Output y = $6.50x
0	$6.50(0) = 0
1	$6.50(1) = $6.50
2	$6.50(2) = $13.00
3	$6.50(3) = $19.50
4	$6.50(4) = $26.00

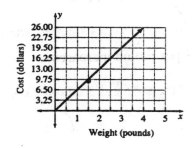

b. From the graph we estimate the cost to be about $16. Using the price per pound we get: $6.50(2.5) = $16.25.

From the graph we estimate the cost to be about $11. Using the price per pound we get: $6.50(1.75) = $11.38.

23. From the graph we estimate the cost to be about $21. Using the price per pound we get: $6.50(3.25) = $21.13.

c. Packaged nuts are a better deal, the data point is below the line on the graph. 1.5 pound of bulk nuts would cost $9.75.

25. a.

Minutes	$ Remaining
0	24 - 1.50(0) = 24
4	24 - 1.50(4) = 18
8	24 - 1.50(8) = 12
12	24 - 1.50(12) = 6
16	24 - 1.50(16) = 0

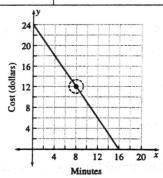

b. $12.00 remains

c. In (0, 24), 24 is the value of the card after talking zero minutes.

d. In (16, 0), 16 is the total minutes of phone time available on the card at $1.50 per minute.

Section 1.4 (con't)

27. a

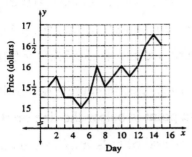

b. Prices are rising slightly.

29. Scale counts by 2's

31. a. 9

b. 12.5

c. 16

d. 30.5

33. $2\frac{1}{2}$ yd.

35. The scale is not the same as on the rest of that axis.

37. 58" needs $1\frac{3}{4}$ yd, 44" needs $2\frac{1}{2}$ yd

$2\frac{1}{2} - 1\frac{3}{4} = \frac{3}{4}$ yd

39. Outputs would be above the 44 inch-wide outputs.

41. (x, y) is called an ordered pair because the order in which the numbers are written is important.

43. The ordered pair is (x, y). Trace vertically to the horizontal axis to find x. Trace horizontally to the vertical axis to find y.

45. The second number is zero on the horizontal axis and the first number is zero on the vertical axis.

47. $20 is always greater than 10% of the input between $20 and $200.

49. a. $20 (greater than 10%)

b. $20 (greater than 10%)

c. $40 [(0.10)$400]

d. $250 [$50+($700-$500)]

51. No, the balance could be anywhere between $20 and $200.

53. The graph becomes steeper when the price per pound increases.

55. A is 2 right of and 3 below (0, -2)

change in $x = 2$, change in $y = -3$

$0 + 2 = 2, \ -2 + (-3) = -5$
$A = (2, -5)$

B is 1 right of and 3 below (2, 3)

change in $x = -2$, change in $y = -3$

$0 + (-2) = -2, \ -2 + (-3) = -5$
$B = (-2, -5)$

Section 1.4 (con't)

57. A is 1 right of and 1 above (2, 3)

change in $x = 1$, change in $y = 1$

$2 + 1 = 3$, $3 + 1 = 4$

$A = (3, 4)$

B is 3 right of and 1 below (2, 3)

change in $x = 3$, change in $y = -1$

$2 + 3 = 5$, $3 + (-1) = 2$

$B = (5, 2)$

Chapter 1 Review

1. Answers will vary - some possibilities are:

   a. You will be done with class and have time to reach the bus stop.

   b. You will write the 8 first and subtract 5.

   c. You will write the 12 first and divide by 8.

   d. Your bicycle is in good working condition.

3. a. 3

   b. 6

   c.

Dots	Dominos
0	1
1	3
2	6
3	10
4	15
5	21
6	28

   d. To find the pattern, look at the change in the # of dominos.

   e. 7 dots = 28 + 8 dominos = 36

   8 dots = 36 + 9 dominos = 45

   9 dots = 45 + 10 dominos = 55

5. a. $\dfrac{5}{6} + \dfrac{3}{8} = \dfrac{20}{24} + \dfrac{9}{24} = \dfrac{29}{24} = 1\dfrac{5}{24}$

   $\dfrac{5}{6} - \dfrac{3}{8} = \dfrac{20}{24} - \dfrac{9}{24} = \dfrac{11}{24}$

   $\dfrac{5}{6} \cdot \dfrac{3}{8} = \dfrac{15}{48} = \dfrac{5}{16}$

   $\dfrac{5}{6} \div \dfrac{3}{8} = \dfrac{5}{6} \cdot \dfrac{8}{3} = \dfrac{40}{18} = \dfrac{20}{9} = 2\dfrac{2}{9}$

   b. $2\dfrac{1}{6} + 1\dfrac{3}{4} = \dfrac{13}{6} + \dfrac{7}{4} = \dfrac{26}{12} + \dfrac{21}{12} =$

   $\dfrac{47}{12} = 3\dfrac{11}{12}$

   $2\dfrac{1}{6} - 1\dfrac{3}{4} = \dfrac{26}{12} - \dfrac{21}{12} = \dfrac{5}{12}$

   $2\dfrac{1}{6} \cdot 1\dfrac{3}{4} = \dfrac{13}{6} \cdot \dfrac{7}{4} = \dfrac{91}{24} = 3\dfrac{19}{24}$

   $2\dfrac{1}{6} \div 1\dfrac{3}{4} = \dfrac{13}{6} \div \dfrac{7}{4} = \dfrac{13}{6} \cdot \dfrac{4}{7} =$

   $\dfrac{52}{42} = \dfrac{26}{21} = 1\dfrac{5}{21}$

7. a. real, rational

   b. real, rational, integer, whole number

   c. real, rational

   d. real, rational

   e. real, rational

   f. real, rational, integer

Chapter 1 Review (con't)

9.

Input	Output
1	$0.89(1) - $0.10 = $0.79
2	$0.89(2) - $0.10 = $1.68
3	$0.89(3) - $0.10 = $2.57
n	0.89n$ - $0.10

11.

Input n	Output $n + 4$
0	$0 + 4 = 4$
1	$1 + 4 = 5$
2	$2 + 4 = 6$
3	$3 + 4 = 7$
4	$4 + 4 = 8$
5	$5 + 4 = 9$
6	$6 + 4 = 10$

13.

Input	Output
0	5
1	$1 - 2 = -1$
2	5
3	$3 - 2 = 1$
4	5
5	$5 - 2 = 3$
6	5

15. a. Four less than 3 times the input.

   b. 3 more than the input times itself.

   c. The input divided by 3.

17. Addition:

increased by; sum; more than; longer than; farther; increases; altogether; plus; combined; faster than; a greater than b; b bigger than a; a exceeds b by 3; $a + b$

Subtraction:

decreased by; fewer than; less than; difference; slower than; loses; a less b; a diminished by b; b subtracted from a; a decreased by b; $a - b$

Multiplication:

product; half, $\frac{1}{2}()$; twice; of; for each; times; one third, $\frac{1}{3}()$; multiplied by; $a \cdot b$; $(a)(b)$; ab; $a(b)$; $a \times b$

Division:

per; half, $\frac{0}{2}$; quotient; one third, $\frac{0}{3}$; the fraction bar; a/b; $a \div b$; b/a

19. $x^2 + 3$ constants are 3 and 1 (implied in $(1)x^2$)

Chapter 1 Review (con't)

21. a. The input is the figure number, the output is the number of circles in the figure.

b. Input 50   Output 54

Input 100   Output 104

c. The output is the input plus 4. $n + 4$

23. a.

Pounds	Cost
x	$\$0.37x$
1	$\$0.37(1) = \$0.37$
2	$\$0.37(2) = \$0.74$
3	$\$0.37(3) = \$1.11$
4	$\$0.37(4) = \$1.48$

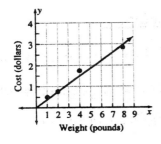

b. $y = \$0.37x$

c. See dots on graph.

d. Answers will vary

25.

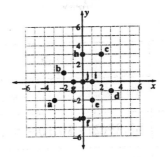

27. A is 5 units right of (2, 3)

$2 + 5 = 7, \ 3 + 0 = 3$

$A = (7, 3)$

B is 3 units below (2, 3)

$2 + 0 = 2, \ 3 + (-3) = 0$

$B = (2, 0)$

29.

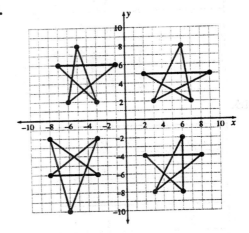

Chapter 1 Review (con't)

31.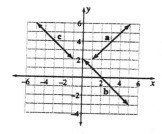

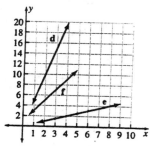

33. a. $y = x + 1;\ (1, 2) = (1, 1 + 1)\,;$

$(2, 3) = (2, 2 + 1)$

b. $y = 2 - x;\ (1, 1) = (1, 2 - 1);$

$(2, 0) = (2, 2 - 2)$

c. $y = 1 - x;\ (-4, 5) = (-4, 1 - (-4));$

$(-3, 4) = (-3, 1 - (-3))$

d. $y = 5x;\ (1, 5) = (1, 5 \cdot 1);$

$(2, 10) = (2, 5 \cdot 2)$

e. $y = \dfrac{x}{2};\ (2, 1) = \left(2, \dfrac{2}{2}\right);$

$(4, 2) = \left(4, \dfrac{4}{2}\right)$

f. $y = 2x + 1;\ (1, 3) = (1, 2 \cdot 1 + 1);$

$(2, 5) = (2, 2 \cdot 2 + 1)$

35.

Input uses	Output cost
0	0
1	0
2	0
3	0
4	0
5	0
6	0
7	0
8	0
9	$(9 - 8)\$0.75 = \$0.75$
10	$(10 - 8)\$0.75 = \$1.50$
11	$(11 - 8)\$0.75 = \$2.25$
12	$(12 - 8)\$0.75 = \$3.00$

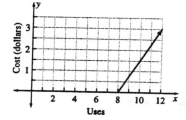

Chapter 1 Test

1. Possible answers:

    The material on the test will be similar to the chapter exercises and chapter review, that the test covers the entire chapter with about the same number of questions from each section.

2. Possible answers: The test has a time limit, students must work individually on the test, the test counts about the same amount toward a grade as other chapter tests.

3. difference

4. set

5. integers

6. **a.** origin

    b. vertical axis or y-axis

    c. horizontal axis or x-axis

    d. Quadrant 3

7. **e.** (-4, 4)

    f. (2, -3)

    g. (0, -5)

    h. (-2, -7)

8.

Input	Output
0	$2(0) + 3 = 3$
1	$2(1) + 3 = 5$
2	$2(2) + 3 = 7$
3	$2(3) + 3 = 9$
4	$2(4) + 3 = 11$
5	$2(5) + 3 = 13$

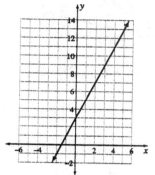

9. **a.** miles on x-axis, cost on y-axis;

    $y = \$0.10x + \$150$

    b. All answers will be in Quadrant 1; inputs given from 0 to 600 by 100's; output above \$150.

    c.

Miles	Cost (\$)
x	$y = 0.10x + 150$
100	$0.10(100) + 150 = 160$
200	$0.10(200) + 150 = 170$
300	$0.10(300) + 150 = 180$
400	$0.10(400) + 150 = 190$
500	$0.10(500) + 150 = 200$
600	$0.10(600) + 150 = 210$

Chapter 1 Test (con't)

9. c.

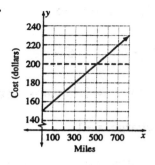

    d. See graph - 500 miles

10.

Input	Output
0	$0 \div 2 = 0$
1	$2(1) = 2$
2	$2 \div 2 = 1$
3	$2(3) = 6$
4	$4 \div 2 = 2$
5	$2(5) = 10$
6	$6 \div 2 = 3$
7	$2(7) = 14$
8	$8 \div 2 = 4$

14.

Input	Output
1	$4(1) = 4$
2	$4(2) = 8$
3	$4(3) = 12$
4	$4(4) = 16$
5	$4(5) = 20$
100	$4(100) = 400$
n	$4n$

The output is 4 times the input.

15. A is 4 to the left and 2 above (-1, 2);

    $-1 + (-4) = -5, \ 2 + 2 = 4$

    $A = (-5, 4)$

    B is 1 to the left and 3 above (-1, 2);

    $-1 + (-1) = -2, \ 2 + 3 = 5$

    $B = (-2, 5)$

11. $\dfrac{5}{8} + \dfrac{1}{6} = \dfrac{15}{24} + \dfrac{4}{24} = \dfrac{19}{24}$

12. $1\dfrac{1}{9} \cdot 1\dfrac{5}{12} = \dfrac{10}{9} \cdot \dfrac{17}{12} = \dfrac{170}{108} = 1\dfrac{31}{54}$

13. $9\% = \dfrac{9}{100}; \ \dfrac{9}{100} \cdot 35 = \dfrac{315}{100} = 3.15$

Section 2.1

1. a. -5

   b. $\frac{1}{2}$

   c. -0.4

   d. $-x$

   e. $2x$

3. a. $|4| = 4$

   b. $|-6| = 6$

   c. $-(-5) = -1 \cdot (-5) = 5$

   d. $-(-2) = -1 \cdot (-2) = 2$

5. a. $-|7| = -1 \cdot |7| = -1 \cdot 7 = -7$

   b. $-|-8| = -1 \cdot |-8| = -1 \cdot 8 = -8$

   c. $-(-3) = -1 \cdot (-3) = 3$

   d. $|-7| = 7$

7. a. $|-4| = 4$

   b. $|5| = 5$

   c. $|4-9| = |-5| = 5$

   d. $-|2+5| = -|7| = -1 \cdot 7 = -7$

9. a. $-(-4) = -1 \cdot (-4) = 4$

   b. $\begin{aligned}-[-(-4)] &= -1 \cdot [-1 \cdot (-4)] = \\ &-1 \cdot 4 = -4\end{aligned}$

11. a. $-|6| = -1 \cdot |6| = -1 \cdot 6 = -6$

   b. $-|-6| = -1 \cdot |-6| = -1 \cdot 6 = -6$

13. The opposite of the absolute value of x.

15. -3 + 4 = 1, net charge = +1

17. -5 + 5 = 0, net charge = 0

19. -7 + 5 = -2, net charge = -2

21. -3 + (+2) = -1,

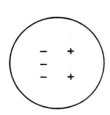

23. +3 + (-5) = -2,

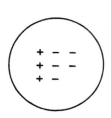

25. a. -8 + 3 = -5

   b. 4 + (-7) = -3

   c. +4 + (-4) = 0

Section 2.1 (con't)

27. a. $-3 + (+3) = 0$

  b. $-4 + (-7) = -11$

  c. $-12 + (+8) = -4$

29. a. $-3 + (-4) + (+5) = -7 + 5 = -2$

  b. $+7 + (-8) + (-3) = 7 + (-11) = -4$

31. a. $2 + (-5) + (-4) = 2 = (-9) = -7$

  b. $-6 + (-3) + (+7) = -9 + 7 = -2$

33. $-12 + 6 - 15 + 7 = -27 + 13 = -14$

35. $-6 + (-7) + (-8) + 20 = -21 + 20 = -1$

37. a. $+1 - (-2) = 3$

  b. $+2 - (+3) = -1$

39. a. $-4 - (-3) = -1$

  b. $-1 - (-3) = +2$

41. a. $8 - 11 = 8 + (-11) = -3$

  b. $-8 - (-11) = -8 + 11 = 3$

  c. $-16 - 3 = -16 + (-3) = -19$

  d. $0 - 5 = 0 + (-5) = -5$

  e. $-17 - 4 = -17 + (-4) = -21$

43. a. $-4 - (-4) = -4 + 4 = 0$

  b. $14 - (10) = 14 + (-10) = 4$

  c $0 - (-5) = 0 + 5 = 5$

  d. $8 - 9 = 8 + (-9) = -1$

  e. $-4 - 5 = -4 + (-5) = -9$

45. a. $2 - 7 = 2 + (-7) = -5$

  b. $2 - 5 = 2 + (-5) = -3$

  c. $2 - (-5) = 2 + 5 = 7$

  d. $-2 - (-5) = -2 + 5 = 3$

  e. $-7 - (-2) = -7 + 2 = -5$

47. a. $8 - (+3) - (-4) =$

     $8 + (-3) + 4 = 5 + 4 = 9$

  b. $-5 - (-7) + (-1) = -5 + 7 + (-1) =$

     $2 + (-1) = 1$

49.

x	y	$x + y$	$x - y$
4	5	$4 + 5 = 9$	$4 - 5 = -1$
5	-4	$5 + (-4) = 1$	$5 - (-4) = 9$
-4	5	$-4 + 5 = 1$	$-4 - 5 = -9$
-4	-5	$-4 + (-5) = -9$	$-4 - (-5) = 1$

51. $6960 \text{ m} - (-40 \text{ m}) = 6960 \text{ m} + 40 \text{ m} = 7000 \text{ m}$

53. $2228 \text{ m} - (-16 \text{ m}) = 2228 \text{ m} + 16 \text{ m} = 2244 \text{ m}$

55. $|-15| + |-10| + 0 + |-5| + |10| =$

     $15 + 10 + 0 + 5 + 10 = 40$

Section 2.1 (con't)

57. $|5| + |12| + 0 + |-7| + |-10| =$

$5 + 12 + 0 + 7 + 10 = 34$

59.

x	$y = |x|$
-3	$|-3| = 3$
-2	$|-2| = 2$
-1	$|-1| = 1$
0	$|0| = 0$
1	$|1| = 1$
2	$|2| = 2$
3	$|3| = 3$

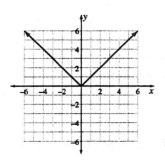

61.

x	$y = |x+1|$
-3	$|-3+1| = |-2| = 2$
-2	$|-2+1| = |-1| = 1$
-1	$|-1+1| = |0| = 0$
0	$|0+1| = |1| = 1$
1	$|1+1| = |2| = 2$
2	$|2+1| = |3| = 3$
3	$|3+1| = |4| = 4$

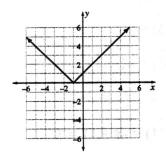

Section 2.1 (con't)

63.

x	$y = \lvert x \rvert - 2$
-3	$\lvert -3 \rvert - 2 = 3 - 2 = 1$
-2	$\lvert -2 \rvert - 2 = 2 - 2 = 0$
-1	$\lvert -1 \rvert - 2 = 1 - 2 = -1$
0	$\lvert 0 \rvert - 2 = 0 - 2 = -2$
1	$\lvert 1 \rvert - 2 = 1 - 2 = -1$
2	$\lvert 2 \rvert - 2 = 2 - 2 = 0$
3	$\lvert 3 \rvert - 2 = 3 - 2 = 1$

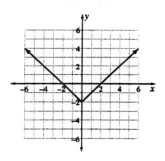

65. True, $\lvert x \rvert = \lvert -x \rvert = x$

67. Absolute value is the distance from
zero on a number line.

69. To subtract, change to adding the
opposite. $a - b = a + (-b)$

Section 2.2

1. $(-2)(+150) = -300$ different signs

3. $(+2)(+400) = +800$ same signs

5. $(-8)(-40) = +320$ different signs

7. $(+3)(-90) = -270$ different signs

9.

Input, x	Output, $3x$
2	$3(2) = 6$
1	$3(1) = 3$
0	$3(0) = 0$
-1	$3(-1) = -3$
-2	$3(-2) = -6$

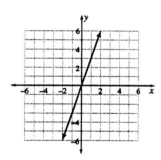

11.

Input, x	Output, $-2x$
2	$-2(2) = -4$
1	$-2(1) = -2$
0	$-2(0) = 0$
-1	$-2(-1) = 2$
-2	$-2(-2) = 4$

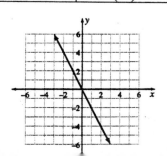

13. a. $(-7)(-6) = 42$ same signs

b. $6(-8) = -48$ different signs

c. $3(-15) = -45$ different signs

d. $(-5)(-7) = 35$ same signs

15. a. $12(-4) = -48$ different signs

b. $8(-8) = -64$ different signs

c. $15(4) = 60$ same signs

d. $(-9)(-9) = 81$ same signs

17. a. $6(-8)(-1) = -48(-1) = 48$

b. $-5(5)(-1) = -25(-1) = 25$

c. $4(3)(-2) = 12(-2) = -24$

19. a. $-5(-6)(-7) = 30(-7) = -210$

b. $-6(9)(2) = -54(2) = -108$

c. $-3(0)(-7) = 0$

21. a. $\dfrac{1}{4}$

b. $-\dfrac{1}{2}$

c. 2

d. $-\dfrac{4}{3}$

e. $0.5 = \frac{1}{2}$, reciprocal $= 2$

Section 2.2 (con't)

23. a. $3\frac{1}{3} = \frac{10}{3}$, reciprocal $= \frac{3}{10}$

  b. $6.5 = 6\frac{1}{2} = \frac{13}{2}$, reciprocal $= \frac{2}{13}$

  c. $\dfrac{1}{x}$

  d. $\dfrac{b}{a}$

  e. $-\dfrac{1}{x}$

25. a. $-3\left(\dfrac{1}{3}\right) = \dfrac{-3}{3} = -1$

  b. $-3\left(\dfrac{-1}{3}\right) = \dfrac{3}{3} = 1$

  c. $2\left(-\dfrac{1}{2}\right) = \dfrac{2}{-2} = -1$

  d. $-\dfrac{1}{4}(-4) = \dfrac{4}{4} = 1$

27. a. The reciprocal of -5 $= -\dfrac{1}{5}$;

    $15\left(-\dfrac{1}{5}\right) = -\dfrac{15}{5} = -3$

  b. The reciprocal of 9 $= \dfrac{1}{9}$;

    $-45 \cdot \dfrac{1}{9} = -\dfrac{45}{9} = -5$

27. c. The reciprocal of -6 $= -\dfrac{1}{6}$;

    $42 \cdot -\dfrac{1}{6} = -\dfrac{42}{6} = -7$

29. a. $\dfrac{56}{-8} = -7$

  b. $\dfrac{-56}{-8} = \dfrac{56}{8} = 7$

  c. $-\dfrac{56}{8} = -7$

31. a. $\dfrac{-55}{-11} = \dfrac{55}{11} = 5$

  b. $\dfrac{-28}{4} = -7$

  c. $\dfrac{-27}{9} = -3$

33.

a	b	$-\left(\dfrac{b}{a}\right)$	$\dfrac{-b}{a}$	$\dfrac{b}{-a}$
5	35	$-\left(\frac{35}{5}\right)$ $=-7$	$\frac{-35}{5}$ $=-7$	$\frac{35}{-5}$ $=-7$
-27	3	$-\left(\frac{3}{-27}\right)$ $=\frac{1}{9}$	$\frac{-3}{-27}$ $=\frac{1}{9}$	$\frac{3}{-(-27)} =$ $\frac{3}{27} = \frac{1}{9}$

Moving the negative sign had no effect on the solution.

35. $\dfrac{-b}{a}$, $-\dfrac{b}{a}$

Section 2.2 (con't)

37.

Input	Output
-3	-1(-3) = 3
-2	-1(-2) = 2
-1	-1(-1) = 1
0	0
1	1
2	2
3	3

39. a. $15 \div \frac{1}{3} = 15 \cdot 3 = 45$

b. $\frac{2}{3} \div \frac{4}{5} = \frac{2}{3} \cdot \frac{5}{4} = \frac{10}{12} = \frac{5}{6}$

c. $\frac{5}{8} \div \frac{2}{3} = \frac{5}{8} \cdot \frac{3}{2} = \frac{15}{16}$

d. $\frac{3}{4} \div \frac{1}{4} = \frac{3}{4} \cdot \frac{4}{1} = \frac{12}{4} = 3$

41.

x	y	xy	$x + y$
2	-2	2(-2) = -4	2 + (-2) = 0
3	-6 ÷ 3 = -2	-6	3 + (-2) = 1
-4	-12 ÷ -4 = 3	-12	-4 + 3 = -1
0 - 3 = -3	3	-3(3) = -9	0
2	-1 - 2 = -3	2(-3) = -6	-1

43. One possible answer is to multiply the absolute values of the numbers.

45. $\dfrac{1}{\frac{2}{3}} = 1 \div \frac{2}{3} = 1 \cdot \frac{3}{2} = \frac{3}{2}$

The answers are the same.

Section 2.3

1. **a.** 2 terms; $2c$ and d

    b. 3 terms; $3x$, $3y$ and -1

    c. 1 term; xy

    d. 2 terms; 1 and $-2a$

    e. 1 term; $-2x$

    f. 2 terms; x^2 and $-y^2$

3. **a.** 3 factors; $\frac{1}{2}$, x and y

    b. 2 factors; $(x + 2)$ and $(x - 2)$

    c. 3 factors; 2, x and $(x + 3)$

    d. 4 factors; 4, a, b and c

    e. 2 factors; $(a + b + c)$ and $\frac{1}{3}$

5. $2\frac{1}{2} + 1\frac{1}{3} + 3\frac{2}{3} + 5\frac{1}{2} =$

    $(2 + 1 + 3 + 5) + (\frac{1}{2} + \frac{1}{2} + \frac{1}{3} + \frac{2}{3}) =$

    $11 + 2 = 13$

7. $\frac{1}{4} \cdot 25 \cdot 8 \cdot 4 = (\frac{1}{4} \cdot 8) \cdot (25 \cdot 4) =$

    $2 \cdot 100 = 200$

9. $4.25 + 2.98 + 1.75 =$

    $(4 + 2 + 1) + (0.25 + 0.75) + 0.98 =$

    $7 + 1 + 0.98 = 8.98$

11. $-3(4)(-5) = -3(-5)(4)$

    $-12(-5) = 15(4)$

    $60 = 60$

Commutative property of multiplication.

13. $-3 + 4 + 5 = 4 + (-3) + 5$

    $1 + 5 = 1 + 5$

    $6 = 6$

Commutative property of addition.

15. $3 + (4 + 5) = (3 + 4) + 5$

    $3 + 9 = 7 + 5$

    $12 = 12$

Associative property of addition.

17. $3(4 \cdot 5) = (3 \cdot 4) + (3 \cdot 5)$

    $3(20) = 12 + 15$

    $60 = 27$

False statement

19. **a.** $8 - (5 - 3) = 8 - 2 = 6$

    b. $(8 - 5) - 3 = 3 - 3 = 0$

    c. $16 \div (4 \div 2) = 16 \div 2 = 8$

    d. $(16 \div 4) \div 2 = 4 \div 2 = 2$

    e. No, subtraction is not associative.

    f. No, division is not associative.

21. $4 \cdot \$4.97 = 4(\$5.00 - \$0.03) =$

    $4 \cdot \$5 - 4 \cdot \$0.03 = \$20 - \$0.12 = \$19.88$

23. $3 \cdot \$10.98 = 3(\$11.00 - \$0.02) =$

    $3 \cdot \$11 - 3 \cdot \$0.02 = \$33 - \$0.06$

    $= \$32.94$

Section 2.3 (con't)

25. a. $6(x+2) = 6x + 6(2) = 6x + 12$

b. $-3(x-3) = -3x + (-3)(-3) =$

$-3x + 9$

c. $-6(x+4) = -6x + (-6)(4) =$

$-6x - 24$

27. a. $-3(x+y-5) =$

$-3x + (-3)y + (-3)(-5) =$

$-3x - 3y + 15$

b. $-(x - y - z) =$

$-1x + (-1)(-y) + (-1)(-z) =$

$-x + y + z$

29. a. $-(x - 3) = -1(x) + (-1)(-3) = -x + 3$

b. $y(4 + y) = 4y + y(y) = 4y + y^2$

c. $-(2 - y) = -1(2) + (-1)(-y) = -2 + y$

31. a. $\dfrac{-2x}{-6x} = \dfrac{-1 \cdot 2 \cdot x}{-1 \cdot 2 \cdot 3 \cdot x} = \dfrac{1}{3}$

b. $\dfrac{-14a}{21a} = \dfrac{-2 \cdot 7 \cdot a}{3 \cdot 7 \cdot a} = -\dfrac{2}{3}$

c. $\dfrac{6x}{15xyz} = \dfrac{2 \cdot 3 \cdot x}{5 \cdot 3 \cdot x \cdot y \cdot z} = \dfrac{2}{5yz}$

33. a. $\dfrac{15xy}{21y} = \dfrac{3 \cdot 5 \cdot x \cdot y}{3 \cdot 7 \cdot y} = \dfrac{5x}{7}$

b. $\dfrac{39abc}{13acd} = \dfrac{3 \cdot 13 \cdot a \cdot b \cdot c}{13 \cdot a \cdot c \cdot d} = \dfrac{3b}{d}$

33. c. $\dfrac{-12xy}{48xz} = \dfrac{-3 \cdot 4 \cdot x \cdot y}{3 \cdot 4 \cdot 4 \cdot x \cdot z} = -\dfrac{y}{4z}$

35. a. $\dfrac{ab}{bc} = \dfrac{a}{c}$

b. $\dfrac{ab}{ac} = \dfrac{b}{c}$

c. $\dfrac{ay}{by} = \dfrac{a}{b}$

37. a. $\dfrac{3x+4}{4} = \dfrac{3x}{4} + \dfrac{4}{4} = \dfrac{3}{4}x + 1$

b. $\dfrac{4x+8}{4} = \dfrac{4x}{4} + \dfrac{8}{4} = x + 2$

c. $\dfrac{x^2 + xy}{x} = \dfrac{x^2}{x} + \dfrac{xy}{x} = x + y$

d. $\dfrac{ab - bc}{b} = \dfrac{ab}{b} - \dfrac{bc}{b} = a - c$

39. $3x^2 + 3x + 6$

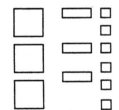

41. $2a^2 + 4ab + 6b^2$

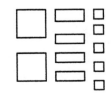

Section 2.3 (con't)

43. a. -4

  b. 1

  c. -1

45. a. $-3x + 4x - 6x + 8x = 3x$

  b. $-6y^2 + 8y^2 - 10y^2 - 3y^2 =$

  $-11y^2$

  c. $2a + 6 + 3a + 9 =$
  $2a + 3a + 6 + 9 = 5a + 15$

  d. $3(x+1) - 2(x-1) =$

  $3x + 3 - 2x + 2 =$
  $3x - 2x + 3 + 2 = x + 5$

  e. $3(x-4) + 4(4-x) =$

  $3x - 12 + 16 - 4x =$
  $3x - 4x - 12 + 16 = -x + 4$

  f. $2x - 3y + 2x - 3y =$

  $2x + 2x - 3y - 3y = 4x - 6y$

47. a. $\frac{1}{2}x + \frac{1}{4}y + \frac{1}{2}y - \frac{1}{4}x =$

  $\frac{1}{2}x - \frac{1}{4}x + \frac{1}{4}y + \frac{1}{2}y =$
  $\frac{1}{4}x + \frac{3}{4}y$

  b. $0.5a + 0.75b - 0.5b + 1.5a =$

  $0.5a + 1.5a + 0.75b - 0.5b$
  $2.0a + 0.25b$

  c. $-2x + 3y + (-4x) + (-6x) =$

  $-2x - 4x - 6x + 3y = -12x + 3y$

47. d. $2(b+c) - 2(b-c) =$

  $2b + 2c - 2b + 2c =$
  $2b - 2b + 2c + 2c = 4c$

49. $x \cdot y = y \cdot x$

51. $a(b + c)$ is the product of two factors, applying the distributive property gives $ab + bc$, the sum of two terms.

53. In like terms variables and exponents are identical.

55. Factors are multiplied; terms are added.

57. $\dfrac{2x + 2y + 2}{2} = \dfrac{2x}{2} + \dfrac{2y}{2} + \dfrac{2}{2} = x + y + 1$;

student left off the 1.

59. a. $45 = 3 \cdot 3 \cdot 5$

  b. 59 is prime.

  c. $72 = 2 \cdot 2 \cdot 2 \cdot 3 \cdot 3$

  d. $111 = 3 \cdot 37$

Mid Chapter 2 Test

1. **a.** $2 - 5 = -3$

   b. $-3 + 5 = 2$

   c. $-3 - (-5) = -3 + 6 = 2$

2. **a.** $3 - (-4) = 3 + 4 = 7$

   b. $-2 + (-5) = -7$

   c. $6 + (-2.5) = 3.5$

3. **a.** $-5.50 + 18.98 - 12.76 =$

   $-5.50 - 12.76 + 18.98 =$

   $-18.26 + 18.98 = 0.72$

   b. $-3.89 - 42.39 + 50.00 =$

   $-46.28 + 50.00 = 3.72$

4. **a.** $(-3)(2) = -6$

   b. $(-5)(-3) = 15$

   c. $-(-4) = (-1)(-4) = 4$

5. **a.** $\dfrac{27}{-3} = \dfrac{3 \cdot 9}{-3} = -9$

   b. $\dfrac{-28}{-2} = \dfrac{-2 \cdot 14}{-2} = 14$

   c. $\dfrac{-32}{4} = \dfrac{-4 \cdot 8}{4} = -8$

6. **a.** $4x + 5y - 2x + y =$

   $4x - 2x + 5y + y = 2x + 6y$

   b. $2x^2 - 3x + 2x(1 - x) =$

   $2x^2 - 3x + 2x - 2x^2 =$

   $2x^2 - 2x^2 - 3x + 2x = -x$

7. **a.** $\dfrac{4xyz}{xy} = 4z$

   b. $\dfrac{3x}{xyz} = \dfrac{3}{yz}$

   c. $\dfrac{-2y}{4xy} = \dfrac{-2y}{2 \cdot 2xy} = -\dfrac{1}{2x}$

8. **a.** $+6 - (+4) = +2$

   b. $0 - (-3) = +3$

9. $-3 + 6 - 24 + 27 =$

   $-3 - 24 + 27 + 6 = -27 + 27 + 6 = 6$

10. $-6 + 12 + 18 - 24 =$

    $-6 - 24 + 12 + 18 =$

    $-30 + 30 = 0$

11. $13 \cdot (-4) \cdot 5 \cdot (-3) =$

    $-3(13) \cdot (-4)(5) = (-39)(-20) = 780$

Mid Chapter 2 Test (con't)

12.

x	y = x - 1
-2	-2 - 1 = -3
-1	-1 - 1 = -2
0	0 - 1 = -1
1	1 - 1 = 0
2	2 - 1 = 1
3	3 - 1 = 2

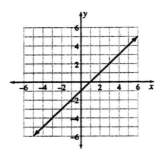

14.

x	y = 2x - 3
-2	2(-2) - 3 = -7
-1	2(-1) - 3 = -5
0	2(0) - 3 = -3
1	2(1) - 3 = -1
2	2(2) - 3 = 1
3	2(3) - 3 = 3

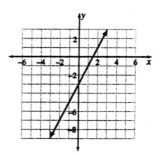

13.

x	y = x + 1
-2	-2 + 1 = -1
-1	-1 + 1 = 0
0	0 + 1 = 1
1	1 + 1 = 2
2	2 + 1 = 3
3	3 + 1 = 4

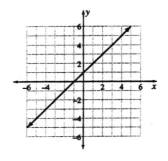

15.

x	y = 3 - x
-2	3 - (-2) = 5
-1	3 - (-1) = 4
0	3 - 0 = 3
1	3 - 1 = 2
2	3 - 2 = 1
3	3 - 3 = 0

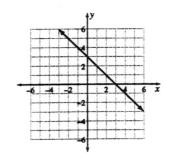

Mid Chapter 2 Test (con't)

16. 8850 m - (-400 m) = 8850 m + 400 m = 9250 m

17. 6194 m - (-86 m) = 6194 m + 86 m = 6280 m

18. Mauna Kea's height from the ocean floor is; 13,710 ft - (-16,400 ft) = 13,710 ft + 16,400 ft = 30,110 ft.

It is the "taller" mountain by: 30,110 ft - 29,028 ft = 1082 ft.

19. $\dfrac{4x + 2y}{2} = \dfrac{4x}{2} + \dfrac{2y}{2} = 2x + y$

20. $\dfrac{ab + bc}{b} = \dfrac{ab}{b} + \dfrac{bc}{b} = a + c$

Section 2.4

1. a. In $3x^2$ the base is x.

$3x^2 = (3)(x)(x);$

3 times the square of x

b. In $-3x^2$ the base is x.

$-3x^2 = (-3)(x)(x);$

The opposite of 3 times the square of x

c. In $(-3x)^2$ the base is $(-3x)$.

$(-3x)^2 = (-3x)(-3x);$

The square of the opposite of $3x$

d. In ax^2 the base is x.

$ax^2 = (a)(x)(x);$

a times x squared

e. In $-x^2$ the base is x.

$-x^2 = (-1)(x)(x);$

The opposite of x squared

f. In $(-x^2)$ the base is $(-x)$.

$(-x)^2 = (-x)(-x);$

The square of the opposite of x

3. a. $3^5 = 3 \cdot 3 \cdot 3 \cdot 3 \cdot 3 = 243$

b. $2^6 = 2 \cdot 2 \cdot 2 \cdot 2 \cdot 2 \cdot 2 = 64$

c. $(-2)^2 = (-2)(-2) = 4$

d. $(-3)^3 = (-3)(-3)(-3) = -27$

5. a. $\left(\frac{1}{3}\right)^3 = \left(\frac{1}{3}\right)\left(\frac{1}{3}\right)\left(\frac{1}{3}\right) = \frac{1}{27}$

b. $\left(\frac{4}{5}\right)^3 = \left(\frac{4}{5}\right)\left(\frac{4}{5}\right)\left(\frac{4}{5}\right) = \frac{64}{125}$

c. $\left(-\frac{2}{3}\right)^3 = \left(-\frac{2}{3}\right)\left(-\frac{2}{3}\right)\left(-\frac{2}{3}\right) = -\frac{8}{27}$

d. $-\left(-\frac{1}{3}\right)^2 = (-1)\left(-\frac{1}{3}\right)\left(-\frac{1}{3}\right) = -\frac{1}{9}$

7. a. $2 \cdot 4^2 = 2 \cdot 4 \cdot 4 = 32$

b. $-2^2 = (-1) \cdot 2 \cdot 2 = -4$

c. $3(-2)^2 = 3(-2)(-2) = 12$

d. $-4 \cdot 3^2 = -4 \cdot 3 \cdot 3 = -36$

9. a. $2^2 2^3 = 2^{2+3} = 2^5 = 32$

b. $\left(3^4\right)^2 = 3^{4 \cdot 2} = 3^8 = 6561$

c. $4^3 4^2 = 4^{3+2} = 4^5 = 1024$

d. $\dfrac{3^7}{3^2} = 3^{7-2} = 3^5 = 243$

11. a. $m^3 m^5 = m^{3+5} = m^8$

b. $n^4 n^4 = n^{4+4} = n^8$

c. $a^6 a^2 = a^{6+2} = a^8$

d. $a^7 a^1 = a^{7+1} = a^8$

Section 2.4 (con't)

13. a. $\dfrac{x^5}{x^2} = x^{5-2} = x^3$

b. $\dfrac{a^8}{a^5} = a^{8-5} = a^3$

c. $\left(\dfrac{x}{y}\right)^2 = \dfrac{x^2}{y^2}$

d. $\left(\dfrac{2x}{y}\right)^3 = \dfrac{(2x)^3}{y^3} = \dfrac{2^3 x^3}{y^3} = \dfrac{8x^3}{y^3}$

15. $8 = 2^3$ and $125 = 5^3$ so n must equal

$3.$ $\left(\dfrac{2}{5}\right)^3 = \dfrac{2^3}{5^3} = \dfrac{8}{125}$

17. The two exponents must add to 12.

Some combinations are:

1, 11; 3, 9; 4, 8; 6, 6;

19. a. $\left(x^2\right)^3 = x^{2\cdot3} = x^6$

b. $(xy)^2 = x^2 y^2$

c. $\left(x^2 y^3\right)^2 = \left(x^2\right)^2\left(y^3\right)^2 =$

$x^{2\cdot2} y^{3\cdot2} = x^4 y^6$

21. a. $(2a)^3(2a)^6 = (2a)^{3+6} =$

$(2a)^9 = 2^9 a^9 = 512a^9$

b. $(2a)^2(3b)^3 = 2^2 a^2 3^3 b^3 =$

$4\cdot27\cdot a^2 b^3 = 108a^2 b^3$

c. $4(3a)^3 (-1)^1 = 4\cdot3^3 \cdot a^3 \cdot(-1) =$

$(-1)\cdot4\cdot27\cdot a^3 = -108a^3$

23. a. $\dfrac{a}{a} = 1$

b. $\dfrac{-4ab}{6a} = \dfrac{-2\cdot2\cdot a\cdot b}{2\cdot3\cdot a} = -\dfrac{2b}{3}$

c. $\dfrac{a^3 b^2}{ab} = a^{3-1} b^{2-1} = a^2 b$

d. $\dfrac{-4x^4 y^3}{-10x^2 y} = \dfrac{-2\cdot2}{-2\cdot5} x^{4-2} y^{3-1} =$

$\dfrac{2}{5} x^2 y^2$

25. a. $\left(x^2\right)^2 = x^{2\cdot2} = x^4$

b. $\left(2x^2\right)^3 = 2^3 x^{2\cdot3} = 8x^6$

c. $\left(-\tfrac{1}{4}a^2\right)^2 = \left(-\tfrac{1}{4}\right)^2 a^{2\cdot2} =$

$\dfrac{(-1)^2}{4^2} a^4 = \tfrac{1}{16} a^4$

Section 2.4 (con't)

27. a. $\left(y^2\right)^3 = y^{2\cdot 3} = y^6$

b. $\left(3y^3\right)^2 = 3^2 y^{3\cdot 2} = 9y^6$

c. $\left(\frac{1}{3}y^3\right)^2 = \left(\frac{1}{3}\right)^2 y^{3\cdot 2} =$

$\frac{1^2}{3^2}y^6 = \frac{1}{9}y^6$

29. a. $1 + 9\cdot 9 - 8 = 1 + 81 - 8 = 74$

b. $1 + \sqrt{9}\cdot 9 - 8 = 1 + 3\cdot 9 - 8 =$

$1 + 27 - 8 = 20$

c. $\left(1 + \sqrt{9}\right)\cdot 9 - 8 = (1 + 3)\cdot 9 - 8 =$

$4\cdot 9 - 8 = 36 - 8 = 28$

31. $2\cdot 3\cdot 4 + 2\cdot 4\cdot 5 + 2\cdot 3\cdot 5 =$

$24 + 40 + 30 = 94$

33. $\frac{1}{2}\cdot 5\cdot (6 + 8) = \frac{1}{2}\cdot 5\cdot 14 =$

$\frac{1}{2}\cdot 14\cdot 5 = 7\cdot 5 = 35$

35. $6^2 + 8^2 = 36 + 64 = 100$

37. $5^2 + 12^2 = 25 + 144 = 169$

39. $7 - 2(4 - 1) = 7 - 2\cdot 3 = 7 - 6 = 1$

41. $(7 - 2)(4 - 1) = 5\cdot 3 = 15$

43. $6 - 3(x - 4) = 6 - 3x + 12 =$

$-3x + 6 + 12 = -3x + 18$

45. $(6 - 3)(x - 4) = 3(x - 4) = 3x - 12$

47. $3\big[8 - 2(3 - 5)\big] = 3\big[8 - 2(-2)\big] =$

$3(8 + 4) = 3\cdot 12 = 36$

49. $|4 - 6| + |6 - 4| = |-2| + |2| = 2 + 2 = 4$

51. $|5 - 2| + |2 - 5| = |3| + |-3| = 3 + 3 = 6$

53. a. $\dfrac{4 - 7}{2 - (-3)} = \dfrac{-3}{2 + 3} = \dfrac{-3}{5}$

b. $\dfrac{-2 - 3}{-2 - (-3)} = \dfrac{-5}{-2 + 3} = \dfrac{-5}{1} = -5$

c. $\dfrac{5 - 2}{3 - (-4)} = \dfrac{3}{3 + 4} = \dfrac{3}{7}$

55. a. $\dfrac{3 + \sqrt{25}}{4} = \dfrac{3 + 5}{4} = \dfrac{8}{4} = 2$

b. $\dfrac{9 - \sqrt{36}}{2} = \dfrac{9 - 6}{2} = \dfrac{3}{2}$

c. $\dfrac{12 + \sqrt{64}}{5} = \dfrac{12 + 8}{5} = \dfrac{20}{5} = 4$

57. $\sqrt{(-6 - 6)^2 + (12 - 3)^2} =$

$\sqrt{(-12)^2 + 9^2} = \sqrt{144 + 81} =$

$\sqrt{225} = 15$

59. $\sqrt{(4 - (-2))^2 + (4 - (-4))^2} =$

$\sqrt{(4 + 2)^2 + (4 + 4)^2} = \sqrt{6^2 + 8^2} =$

$\sqrt{36 + 64} = \sqrt{100} = 10$

Section 2.4 (con't)

61. $\dfrac{|3-7.5|+|9-7.5|+|10-7.5|}{3} =$

$\dfrac{|-4.5|+|1.5|+|2.5|}{3} =$

$\dfrac{4.5+1.5+2.5}{3} = \dfrac{8.5}{3} \approx 2.83$

63. $3.14(2)^2 \cdot 6 = 3.14 \cdot 4 \cdot 6 = 75.36$

65. $\frac{4}{3}(3.14) \cdot 2^3 = \frac{4}{3}(3.14) \cdot 8 =$

$\dfrac{4 \cdot 3.14 \cdot 8}{3} = \dfrac{100.48}{3} \approx 33.49$

67. $2(3.14) \cdot 3 \cdot 4 + 2(3.14) \cdot 3^2 =$

$6.28 \cdot 12 + 6.28 \cdot 9 \approx 131.88$

69. $-\frac{1}{2}(32)(-1)^2 + 16(-1) + 50 =$

$-16 + (-16) + 50 = 18$

71. One answer is to complete all arithmetic possible in an expression.

73. a. When multiplying expressions with like bases, keep the base and add the exponents.

  b. When applying an exponent to a power expression, multiply the exponents.

  c. When applying an exponent to a product in parenthesis, apply the exponent to all parts of the product.

73. d. When applying an exponent to a quotient in parenthesis, apply the exponent to all parts of the quotient.

75. Student #1 squared the exponent on x.

Student #2 multiplied the numerical coefficient 3 by the exponent.

Student #3 is correct.

77. a. $1 \cdot 9 + 9 - 8 = 10$? true because

$9 + 9 - 8 = 18 - 8 = 10$

$1 - 9 + 9 + 8 = 9$? true because

$-8 + 9 + 8 = 1 + 8 = 9$

$1[9 - (9 - 8)] = 8$? true because

$9 - (9 - 8) = 9 - 1 = 8$

$-1 + 9 - 9 - 8 = 7$? false because

$8 - 9 - 8 = -1 - 8 = -9$

Should be $-1 + 9 - 9 \underline{+ 8} = 7$

$-1 - 9 \div 9 + 8 = 6$? true because

$-1 - (\frac{9}{9}) + 8 = -1 - 1 + 8 = 6$

  b. $-1 + (9 + 9) \div 9 = 1$? true because

$-1 + 18 \div 9 = -1 + 2 = 1$

$1 \cdot (9 + 9) \div 9 = 2$? true because

$18 \div 9 = 2$

$1 + (9 + 9) \div 9 = 3$? true because

$1 + 18 \div 9 = 1 + 2 = 3$

Section 2.4 (con't)

77. b. $1 \cdot (9 \div 9) + \sqrt{9} = 4$? true because

$1 \cdot 1 + 3 = 1 + 3 = 4$

$1 + (9 \div 9) - \sqrt{9} = 5$? false

$1 + 1 - 3 = -1$

Should be:

$1 + (9 \div 9) \underline{+} \sqrt{9} = 5$

79. When using the distributive property on $-3(x - 1)$ the student got an incorrect sign for $(-3)(-1)$. Should be $+3$.

81.

# of Game days	Teams remaining
0	64
1	$64 \cdot \frac{1}{2} = 32$
2	$64 \cdot \frac{1}{2} \cdot \frac{1}{2} =$ $64\left(\frac{1}{2}\right)^2 = 16$
n	$64\left(\frac{1}{2}\right)^n$

Final game has 2 teams: $2 = 64\left(\frac{1}{2}\right)^n$

Continue the table pattern to get 5 game days with 2 teams remaining. The final game is played on the 6th game day which is a Sunday.

83.

# of Rnds	Money returned	# of Nickels returned
1	$0.75 \cdot \$20$ $= \$15$	$0.75 \cdot 400 =$ 300
2	$0.75(0.75 \cdot \$20) =$ $(0.75)^2 \cdot \$20 =$ $\$11.25$	$0.75(0.75 \cdot 400) =$ $(0.75)^2 \cdot 400 =$ 225
n	$(0.75)^n \cdot \$20$	$(0.75)^n \cdot 400$

Player is through when;

$(0.75)^n(\$20) < \$0.05$ or $(0.75)^n(400) < 1$.

Using guess & check: $n = 21$

$(0.75)^{21}(20) \approx 0.0024(20) \approx 0.048$

$(0.75)^{21}(400) \approx 0.0024(400) \approx 0.96$

Section 2.5

1. $\dfrac{8\,oz \cdot 1\,lb}{16\,oz} = \dfrac{1}{2}\,lb$

3. $\dfrac{100\,yd \cdot 3\,ft \cdot 12\,in}{1\,yd \cdot 1\,ft} = 100 \cdot 3 \cdot 12\,in =$

3600 in

5. $\dfrac{feet^3}{foot} = \dfrac{foot \cdot feet^2}{foot} = feet^2$

7. $\dfrac{inches^2}{inches^3} = \dfrac{inches^2}{inches^2 \cdot inch} = \dfrac{1}{inch}$

9. $\dfrac{1\,km}{1} \cdot \dfrac{1000\,m}{1\,km} \cdot \dfrac{39.37\,in}{1\,m} \cdot \dfrac{1ft}{12\,in} \approx$

3281 ft.

11. $\dfrac{120\,lb}{1} \cdot \dfrac{1\,kg}{2.2\,lb} \cdot \dfrac{1000\,g}{1\,kg} \approx 54{,}545\,g$

13. $\dfrac{1\,day}{1} \cdot \dfrac{24\,hrs}{1\,day} \cdot \dfrac{60\,min}{1\,hr} \cdot \dfrac{60\,sec}{1\,min} =$

86,400 sec

15. $\dfrac{1{,}000{,}000\,sec}{1} \cdot \dfrac{1\,min}{60\,sec} \cdot \dfrac{1\,hr}{60\,min} \cdot$

$\dfrac{1\,day}{24\,hr} \approx 11.6\,days$

17. From Exercise 13,

$\dfrac{72\,yr}{1} \cdot \dfrac{365\,days}{1\,yr} \cdot \dfrac{86{,}400\,sec}{1\,day} =$

2,270,592,000 sec

19. $\dfrac{1\,ft^3}{1} \cdot \dfrac{12\,in}{1\,ft} \cdot \dfrac{12\,in}{1\,ft} \cdot \dfrac{12\,in}{1\,ft} = 1728\,in^3$

21. $\dfrac{1\,yd^2}{1} \cdot \dfrac{36\,in}{1\,yd} \cdot \dfrac{36\,in}{1\,yd} = 1296\,in^2$

23. $\dfrac{1200\,in^2}{1} \cdot \dfrac{1\,ft}{12\,in} \cdot \dfrac{1\,ft}{12\,in} = \dfrac{1200\,ft^2}{144} \approx$

8.3 ft^2

25. $\dfrac{19.77\,km}{1} \cdot \dfrac{3281\,ft}{1\,km} \approx 65{,}000\,ft$

27. a. $P = 1.8\,cm + 2.0\,cm + 2.1\,cm +$

$4.0\,cm = 9.9\,cm$

$A = \frac{1}{2}(1.6\,cm)(2.0\,cm + 4.0\,cm)$

$= (0.8)(6.0)\,cm^2 = 4.8\,cm^2$

b. $P = 2.0\,cm + 3.0\,cm + 2.1\,cm +$

$5.5\,cm = 12.6\,cm$

$A = \frac{1}{2}(1.6\,cm)(3.0\,cm + 5.5\,cm)$

$= (0.8)(8.5)\,cm^2 = 6.8\,cm^2$

c. Perimeter is circumference;

$C = 2\pi(2\,ft) = 4\pi\,ft \approx 12.6\,ft$

$A = \pi(2\,ft)^2 = 4\pi\,ft^2 \approx 12.6\,ft^2$

Section 2.5 (con't)

27. d. $P = 3.5 \text{ cm} + 2.3 \text{ cm} + 3.8 \text{ cm} =$

9.6 cm

$A = \frac{1}{2}(3.8 \text{ cm})(2.1 \text{ cm})$

$= (1.9)(2.1) \text{ cm}^2 \approx 4.0 \text{ cm}^2$

29. a. $P = 26 \text{ yd} + 10 \text{ yd} + 24 \text{ yd} =$

60 yd

$A = \frac{1}{2}(24 \text{ yd})(10 \text{ yd})$

$= (12)(10) \text{ yd}^2 = 120 \text{ yd}^2$

b. $P = (4)(1.7 \text{ cm}) = 6.8 \text{ cm}$

$A = (1.7 \text{ cm})^2 \approx 2.9 \text{ cm}^2$

c. $P = 31.9 \text{ m} + 17 \text{ m} + 36.1 \text{ m} =$

85 m

$A = \frac{1}{2}(36.1 \text{ m})(15 \text{ m}) \approx 270.8 \text{ m}^2$

d. Perimeter is circumference

$C = \pi(9 \text{ in}) = 9\pi \text{ in} \approx 28.3 \text{ in}$

$A = \pi(\frac{9}{2} \text{ in})^2 = \frac{81}{4}\pi \text{ in}^2 \approx 63.6 \text{ in}^2$

31. a. $S = 2(10 \text{ ft})(5 \text{ ft}) + 2(6 \text{ ft})(10 \text{ ft})$

$+ 2(6 \text{ ft})(5 \text{ ft}) = (100 + 120 + 60) \text{ ft}^2$

$= 280 \text{ ft}^2$

$V = (10 \text{ ft})(6 \text{ ft})(5 \text{ ft}) = 300 \text{ ft}^3$

31. b. $S = 2(12 \text{ in})(5 \text{ in}) + 2(4 \text{ in})(12 \text{ in})$

$+ 2(5 \text{ in})(4 \text{ in})$

$= (120 + 96 + 40) \text{ in}^2 = 256 \text{ in}^2$

$V = (12 \text{ in})(5 \text{ in})(4 \text{ in}) = 240 \text{ in}^3$

33. a. $S = 2\pi(6 \text{ cm})^2 + 2\pi(6 \text{ cm})(8 \text{ cm}) =$

$(72\pi + 96\pi) \text{ cm}^2 = 168\pi \text{ cm}^2 \approx$

527.5 cm^2

$V = \pi(6 \text{ cm})^2(8 \text{ cm}) = 288\pi \text{ cm}^3 \approx$

904.3 cm^3

b. $S = 2\pi(3 \text{ cm})^2 + 2\pi(3 \text{ cm})(4 \text{ cm}) =$

$(18\pi + 24\pi) \text{ cm}^2 = 42\pi \text{ cm}^2 \approx$

131.9 cm^2

$V = \pi(3 \text{ cm})^2(4 \text{ cm}) = 36\pi \text{ cm}^3 \approx$

113.0 cm^3

35. a. $S = 4\pi(8 \text{ cm})^2 = 256\pi \text{ cm}^2 \approx$

803.8 cm^2

$V = \frac{4}{3}\pi(8 \text{ cm})^3 = \frac{2048}{3}\pi \text{ cm}^3 \approx$

2143.6 cm^3

b. $S = 4\pi(4 \text{ cm})^2 = 64\pi \text{ cm}^2 \approx$

201.0 cm^2

$V = \frac{4}{3}\pi(4 \text{ cm})^3 = \frac{256}{3}\pi \text{ cm}^3 \approx$

267.9 cm^3

Section 2.5 (con't)

37. The area of a 9" pizza is:

$$\pi(9 \text{ in})^2 = 81\pi \text{ in}^2$$

The area of a 6" pizza is:

$$\pi(6 \text{ in})^2 = 36\pi \text{ in}^2$$

$$\frac{81\pi \text{ in}^2}{36\pi \text{ in}^2} = \frac{81}{36} = 2.25$$

39. $V = (4.5 \text{ in})(7.0 \text{ in})(13.5 \text{ in})$

$$\approx 425.25 \text{ in}^3$$

$$(1000)(425.25) \text{ in}^3 = 425,250 \text{ in}^3$$

$$425,250 \text{ in}^3 \cdot \left(\frac{1 \text{ ft}}{12 \text{ in}}\right)^3 =$$

$$\frac{425,250 \text{ in}^3 \cdot 1 \text{ ft}^3}{1728 \text{ in}^3} \approx 246.1 \text{ ft}^3$$

41. $\dfrac{12 \text{ oz}}{1 \text{ spill}} \cdot \dfrac{1 \text{ cup}}{8 \text{ oz}} \cdot \dfrac{1 \text{ qt}}{4 \text{ cups}} \cdot \dfrac{1 \text{ gal}}{4 \text{ qt}} \cdot$

$$\frac{231 \text{ in}^3}{1 \text{ gal}} \cdot \frac{1 \text{ spill}}{\frac{1}{16} \text{ in}} = 346.5 \text{ in}^2$$

43. $200 \text{ serv.} \cdot \dfrac{12 \text{ oz}}{1 \text{ serv}} \cdot \dfrac{1 \text{ cup}}{8 \text{ oz}} \cdot \dfrac{1 \text{ qt}}{4 \text{ cup}} \cdot$

$$\frac{1 \text{ gal}}{4 \text{ qt}} \approx 19 \text{ gal}$$

45. $\dfrac{16 \text{ oz}}{1 \text{ box}} \cdot \dfrac{1 \text{ box}}{32 \text{ serv}} \cdot \dfrac{1 \text{ serv}}{12 \text{ crackers}} \cdot$

$$\frac{140 \text{ cal}}{1 \text{ oz}} \approx \frac{5.8 \text{ cal}}{1 \text{ cracker}}$$

47. $2,500,000 \text{ gal} \cdot \dfrac{231 \text{ in}^3}{1 \text{ gal}} \cdot \dfrac{1 \text{ ft}^3}{12^3 \text{ in}^3} \cdot$

$$\frac{1 \text{ width}}{40 \text{ ft}} \cdot \frac{1 \text{ thickness}}{6 \text{ ft}} \approx 1392.5 \text{ ft}$$

49. $2,500,000 \text{ gal} \cdot \dfrac{231 \text{ in}^3}{1 \text{ gal}} \cdot \dfrac{1 \text{ ft}^3}{12^3 \text{ in}^3} \cdot$

$$\frac{1 \text{ depth}}{1 \text{ in}} \approx 4,010,417 \text{ ft}^2 \approx 0.14 \text{ mi}^2$$

51. a. $P = 2x \cdot 4 = 8x; \ A = (2x)^2 = 4x^2$

b. $P = 4x \cdot 4 = 16x; \ A = (4x)^2 = 16x^2$

$$\frac{8x}{16x} = \frac{1}{2} \qquad \frac{4x^2}{16x^2} = \frac{1}{4}$$

53. Multiply the base times the height and divide by 2.

55. Multiply the cube of the radius by $\frac{4}{3}\pi$.

57. Both describe regions to be covered in square units.

59. Add 2 lengths and 2 widths.

61. Blue area is a 3 cm square minus a 3 cm diameter circle.

$$A = (3 \text{ cm})^2 - \pi(\tfrac{3}{2} \text{ cm})^2 = (9 - \tfrac{9}{4}\pi) \text{ cm}^2$$

$$\approx 1.9 \text{ cm}^2$$

Section 2.5 (con't)

63. Inside area is a 100 m by 73 m
rectangle plus 2 half circles of 73 m
diameter.

$A = (100 \text{ m})(73 \text{ m}) + \pi(\frac{73}{2} \text{ m})^2 =$

$(7300 + \frac{5329}{4}\pi) \text{ m}^2 \approx 11{,}483.3 \text{ m}^2$

65. Shaded area is a 15 m by 9 m
rectangle minus a (15-12) m by (9-3)
m rectangle.

$A = (15 \text{ m})(9 \text{ m}) - (3 \text{ m})(6 \text{ m}) =$

$(135 - 18) \text{ m}^2 = 117 \text{ m}^2$

Section 2.6

1.

3. **a.** -8 < -3

   b. +4 > -9

   c. $(-3)^2 = 3^2$

   d. $0.5 > 0.5^2$

   e. 6 > -5

   f. $(-2)(6) < (-2)(-5)$

   $-12 < 10$

   g. -6 < -5

   h. $(-2)(-6) > (-2)(-5)$

   $12 > 10$

5. **a.** -3.75 < -3.25

   b. 3(-2) = -3(2)

   c. $\frac{1}{2} > -\frac{1}{2}$

   d. $|-4| > |2|$

   $4 > 2$

   e. $(-2)(-3) > 2(-4)$

   $6 > -8$

   f. $\left(\frac{1}{2}\right)^2 = \left(-\frac{1}{2}\right)^2$

   g. -2.5 > -3

   h. $\frac{22}{7} > \pi$

7. **a.** $0 < x < 4$

   b. $-5 < x \le -2$

9. **a.** $x > 3$ and $x < 8$

   b. $x > -3$ and $x \le -1$

   c. $x > -2$ and $x < 1$

11. f, s; $x < 3$, $(-\infty, 3)$

13. e, w; $x \ge 3$, $[3, +\infty)$

15. b, r; $-3 \le x \le 3$, $[-3, 3]$

17. **a.** $[-1, 3)$; x is between -1 and 3, including -1;

   b. $-4 < x \le -1$; x is greater than -4 and less than or equal to -1;

   c. $-3 \le x < 5$; $[-3, 5)$;

   d. $(-\infty, -4)$; x is less than -4;

   e. $-2 \le x < 5$; $[-2, 5)$; x is between -2 and 5, including -2

Section 2.6 (con't)

17. f. $(-2, +\infty)$; x is greater than -2;

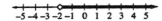

    g. $-4 < x \le 2$; $(-4, 2]$;

    h. $x \ge -3$; x is greater than or equal to -3;

19. A statement that one quantity is greater than or less than another quantity.

21. An interval is a set of numbers between endpoints that may also include one or both endpoints.

23. x cannot be less than 2 <u>and</u> greater than 4.

25. 2 is equal to 2 and $\le$ includes the possibility of equality.

27. a. $x \le 2000$

    b. $2000 < x \le 5000$

    c. $x > 5000$

29. a. $x \le -5$

    b. $-5 < x < 5$

    c. $x \ge 5$

31. a. $x < 5$

    b. $5 \le x \le 50$

    c. $x > 50$

33.

$n\%$	$1.00	$5.00	$10.00	x
6	$0.06(1) = \$0.06$	$0.06(5) = \$0.30$	$0.06(10) = \$0.60$	$0.06x$
10	$0.10(1) = \$0.10$	$0.10(5) = \$0.50$	$0.10(10) = \$1.00$	$0.10x$
25	$0.25(1) = \$0.25$	$0.25(5) = \$1.25$	$0.25(10) = \$2.50$	$0.25x$
100	$1(1) = \$1.00$	$1(5) = \$5.00$	$1(10) = \$10.00$	x
150	$1.5(1) = \$1.50$	$1.5(5) = \$7.50$	$1.5(10) = \$15.00$	$1.5x$

Chapter 2 Review

1. a. $-3 + (-5) = -8$

    b. $3 + (-8) = -5$

    c. $4 - 17 = 4 + (-17) = -13$

    d. $-5 - (-18) = -5 + (+18) = 13$

    e. $-21 + 7 = -14$

    f. $-26 + 19 = -7$

    g. $14 - (-28) = 14 + (+28) = 42$

    h. $12 - 36 = 12 + (-36) = -24$

    i. $-32 - (-16) = -32 + (+16) = -16$

    j. $-4 - (-16) = -4 + (+16) = 12$

    k. $-11 + 22 = 11$

    l. $8 - (-5) = 8 + (+5) = 13$

3. a. $(-9)(6) = -54$

    b. $(-9)(-6) = 54$

    c. $(-18)(-3) = 54$

    d. $(-18)(3) = -54$

    e. $(-8)(-7) = 56$

    f. $(8)(-7) = -56$

    g. $4(-14) = -56$

    h. $(-4)(-14) = 56$

    i. $(-48) \div (-24) = 2$

    j. $(-48) \div 12 = -4$

    k. $48 \div (-6) = -8$

3. l. $(-48) \div (-6) = 8$

    m. $(-48) \div (-3) = 16$

    n. $(-48) \div 3 = -16$

    o. $48 \div (-8) = -6$

5. a. $-(-2)^2 = -4$

    b. $4 - (-2) + (-2)^2 = 4 + 2 + 4 = 10$

    c. $5 - (-3) + (-2)^2 = 5 + 3 + 4 = 12$

    d. $-(-3)^2 = -9$

    e. $\sqrt{(3^2 + 4^2)} = \sqrt{9 + 16} = \sqrt{25} = 5$

    f. $\sqrt{(8^2 + 6^2)} = \sqrt{64 + 36} = \sqrt{100} = 10$

    g. $\sqrt{(25^2 - 20^2)} = \sqrt{625 - 400} = \sqrt{225}$
        $= 15$

    h. $\sqrt{(15^2 - 12^2)} = \sqrt{225 - 144} = \sqrt{81}$
        $= 9$

    i. $\sqrt{(1.5^2 + 2^2)} = \sqrt{2.25 + 4} = \sqrt{6.25}$
        $= 2.5$

    j. $\sqrt{(10^2 - 6^2)} = \sqrt{100 - 36} = \sqrt{64}$
        $= 8$

    k. $-|-4| = -4$

    l. $|-6 - (-5)| = |-6 + 5| = |-1| = 1$

Chapter 2 Review (con't)

7. a. $2x + 2y = 2(x + y)$

b. $ac + ab = a(c + b)$

c. $4x^2 - 8x + 12 = 4(x^2 - 2 + 3)$

d. $3xy + 4x^2y = xy(3 + 4x)$

e. $6x + 12y - 15 = 3(2x + 4y - 5)$

f. $15a^2bc + 5ab^2 + 10abc =$

$5ab(3ac + b + 2c)$

9. a. $\dfrac{abc}{bcd} = \dfrac{a}{d} \cdot \dfrac{bc}{bc} = \dfrac{a}{d}$

b. $\dfrac{4xy}{6xz} = \dfrac{2 \cdot 2y}{2 \cdot 3z} = \dfrac{2y}{3z}$

c. $\dfrac{-21cd}{14ad} = \dfrac{-7 \cdot 3cd}{7 \cdot 2a} = -\dfrac{3c}{2a}$

d. $(-2x)^2 = (-2)^2x^2 = 4x^2$

e. $(-3y)^3 = (-3)^3y^3 = -27y^3$

f. $(-2y)^4 = (-2)^4y^4 = 16y^4$

g. $(-ab)^2 = (-a)^2b^2 = a^2b^2$

h. $(ab)^2 = a^2b^2$

i. $m^4m^5 = m^{4+5} = m^9$

j. $m^2m^7 = m^{2+7} = m^9$

k. $m^5 \div m^2 = m^{5-2} = m^3$

l. $m^7 \div m^4 = m^{7-4} = m^3$

9. m. $\dfrac{3x + 6y}{3} = \dfrac{3x}{3} + \dfrac{6y}{3} = x + 2y$

n. $\dfrac{mn + n^2}{n} = \dfrac{mn}{n} + \dfrac{n^2}{n} = m + n$

o. $\dfrac{2a + 4b}{4} = \dfrac{2a}{4} + \dfrac{4b}{b} = \tfrac{1}{2}a + b$

11. a. $4 - 3(3 - 5) = 4 - 3(-2) = 4 + 6 = 10$

b. $(4 - 3)(3 - 5) = 1(-2) = -2$

c. $\sqrt{5^2 - 4(2)(-12)} = \sqrt{25 + 96} =$

$\sqrt{121} = 11$

d. $\dfrac{4 - \sqrt{49}}{4} = \dfrac{4 - 7}{4} = -\dfrac{3}{4}$

e. $|6 - 1| - |3 - 9| = |5| - |-6| = 5 - 6 = -1$

f. $(7 - 2)^2 + (4 - 1)^2 = 5^2 + 3^2 =$

$25 + 9 = 34$

g. $\dfrac{-3 - (-5)}{-6 - 4} = \dfrac{-3 + 5}{-10} = \dfrac{2}{-10} = -\dfrac{1}{5}$

h. $\tfrac{1}{2} \cdot 11(5 + 7) = \tfrac{1}{2} \cdot 11 \cdot 12 = \tfrac{12}{2} \cdot 11 =$

$6 \cdot 11 = 66$

13. a. $A = \pi(2.5 \text{ ft})^2 = \pi(6.25) \text{ ft}^2 \approx 19.6 \text{ ft}^2$

b. $A = \tfrac{1}{2}(5 \text{ yd})(4 \text{ yd}) = 10 \text{ yd}^2$

c. $V = \tfrac{4}{3}\pi(3 \text{ m})^3 = 4\pi(9) \text{ m}^3 \approx$

113.1 m^3

d. $V = (1.5 \text{ cm})^3 \approx 3.4 \text{ cm}^3$

Chapter 2 Review (con't)

15. a $P = (2.0 + 1.7 + 2.0 + 3.8)\ \text{cm}$

   $= 9.5\ \text{cm}$

   $A = \frac{1}{2}(1.5\ \text{cm})(1.7 + 3.8)\ \text{cm}$

   $= 0.75(5.5)\ \text{cm}^2 \approx 4.1\ \text{cm}^2$

b. $P = 6(1.5\ \text{ft}) = 9\ \text{ft}$

   $A = (1.5\ \text{ft})(2.6\ \text{ft})$

   $+ \frac{1}{2}(2.6\ \text{ft})(0.75\ \text{ft})(2) =$

   $(3.9 + 1.95)\ \text{ft}^2 \approx 5.9\ \text{ft}^2$

c. $P = 2(4\ \text{in}) + 2(5\ \text{in}) = (8 + 10)\ \text{in} =$
18 in

   $A = (5\ \text{in})(3\ \text{in}) = 15\ \text{in}^2$

d. $P = 3.14(4\ \text{m}) + 2(4\ \text{m})$

   $= (12.56 + 8)\ \text{m} \approx 20.6\ \text{m}$

   $A = (4\ \text{m})(4\ \text{m}) + 3.14(2\ \text{m})^2$

   $= (16 + 12.56)\ \text{m}^2 \approx 28.6\ \text{m}^2$

17. One possible example:

   $(3 + 4) + 5 = 7 + 5 = 12:$

   $3 + (4 + 5) = 3 + 9 = 12$

19. The small circle excludes the point and
   is used with $<$ and $>$. The dot includes
   the point and is used with $\leq$ and $\geq$.

21. $-x^2$ is the opposite of $(x)(x)$ while $(-x)^2$ is

   $(-x)(-x).$

23. Change $2\frac{1}{4}$ to $\frac{9}{4}$ and write $\frac{4}{9}$.

29. a. $4 > -3$

   b. $2(-3) < (-2)(-3),\ -6 < 6$

   c. $(-2)^2 > -2^2,\ 4 > -4$

   d. $|-4| = |4|,\ 4 = 4$

   e. $-2^3 = (-2)^3,\ -8 = -8$

   f. $|-5| > -|5|,\ 5 > -5$

   g. $-\frac{1}{4} > -\frac{1}{2}$

   h. $-1.3 > -1.5$

31.

Input: Cost of item	Inequality	Interval
Less than $50	$0 \leq x < \$50$	$[0, \$50)$
$50 to $500	$\$50 \leq x \leq \$500$	$[\$50, \$500]$
Over $500	$x > \$500$	$(\$500, +\infty)$

Chapter 2 Test

1.

x	$y = 3 - x$
-2	3 - (-2) = 5
0	3 - 0 = 3
2	3 - 2 = 1
4	3 - 4 = -1

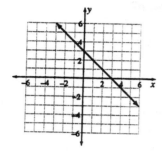

2. a. $\dfrac{3}{4} + \dfrac{5}{6} = \dfrac{9}{12} + \dfrac{10}{12} = \dfrac{19}{12}$

b. $\dfrac{3}{4} - \dfrac{5}{6} = \dfrac{9}{12} - \dfrac{10}{12} = -\dfrac{1}{12}$

c. $\dfrac{3}{4} \cdot \dfrac{-5}{6} = \dfrac{3(-5)}{4(6)} = \dfrac{-15}{24} = \dfrac{-5}{8}$

d. $\dfrac{-3}{4} \div \dfrac{5}{6} = \dfrac{-3}{4} \cdot \dfrac{6}{5} = \dfrac{(-3)(6)}{(4)(5)} =$

$\dfrac{-18}{20} = \dfrac{-9}{10}$

3. Commutative property of addition.

4. Associative property of addition.

5. a. $-5 + 9 = 4$

b. $-1.4 + 2.5 - 3.6 = 2.5 - 5.0 = -2.5$

c. $-4 - (-3) = -4 + 3 = -1$

5. d. $(-3)(4)(-5) = -12(-5) = 60$

e. $8 - (-3)^2 = 8 - 9 = -1$

f. $\sqrt{26^2 - 24^2} = \sqrt{676 - 576} =$

$\sqrt{100} = 10$

g. $m^2 m^9 = m^{2+9} = m^{11}$

h. $m^7 \div m^3 = m^{7-3} = m^4$

i. $36 \div 2 \cdot 2 - 3 + (3^2 - 5) =$

$36 \div 2 \cdot 2 - 3 + (9 - 5) =$
$36 \div 2 \cdot 2 - 3 + 4 = 18 \cdot 2 - 3 + 4 =$
$36 - 3 + 4 = 37$

j. $\dfrac{ace}{aft} = \dfrac{a}{a} \cdot \dfrac{ce}{ft} = \dfrac{ce}{ft}$

k. $\dfrac{6x^2}{9x} = \dfrac{3 \cdot 2}{3 \cdot 3} x^{2-1} = \tfrac{2}{3}x$

l. $\dfrac{-x^3}{(-x)^2} = \dfrac{-x^3}{x^2} = -x^{3-2} = -x$

m. $(a^2 b^3)^3 = a^{2\cdot3} b^{3\cdot3} = a^6 b^9$

n. $\left(\dfrac{a^3}{2b^2}\right)^3 = \dfrac{a^{3\cdot3}}{2^3 b^{2\cdot3}} = \dfrac{a^9}{8b^6}$

o. $\dfrac{3x - 9}{3} = \dfrac{3x}{3} - \dfrac{9}{3} = x - 3$

p. $\dfrac{x^2 + 2x}{x} = \dfrac{x^2}{x} + \dfrac{2x}{x} = x + 2$

Chapter 2 Test (con't)

6. a. $3x + 2y - 2x - 3y + 4x - 4y =$

$3x - 2x + 4x + 2y - 3y - 4y =$

$5x - 5y$

b. $x^3 - 3x^2 + x - 2x^2 + 6x - 2 =$

$x^3 - 3x^2 - 2x^2 + x + 6x - 2 =$

$x^3 - 5x^2 + 7x - 2$

c. $2(x - 2) + 3(x - 1) =$

$2x - 4 + 3x - 3 = 5x - 7$

d. $12(x - 1) - 5(x - 1) =$

$7(x - 1) = 7x - 7$

e. $3a^2 + ab + 6b^2$

7. a. $3(2x + 9y) = 6x + 27y$

b. $3(2a + 9b) = 6a + 27b$

c. $2(3x^2 + 4x - 2) = 6x^2 + 8x - 4$

d. $ab(b - ab + a^2) =$

$ab^2 - a^2b^2 + a^3b$

8. No. $(-x)^2$ is $(-x)(-x)$ while $-x^2$ is the opposite of $x \cdot x$.

9. a. $\pi(2^2) \approx 3.14(4) \approx 12.56 \text{ ft}^2$

b. $\pi(20^2) \approx 3.14(400) \approx 1256 \text{ ft}^2$

c. 100 times as large; 100

10. $5,616,000 \sec \cdot \dfrac{1 \min}{60 \sec} \cdot \dfrac{1 \text{ hr}}{60 \min} \cdot$

$\dfrac{1 \text{ day}}{24 \text{ hr}} = 65 \text{ days}$

11. a. $18,446,400 \text{ in}^2 \cdot \left(\dfrac{1 \text{ ft}}{12 \text{ in}}\right)^2 =$

$128,100 \text{ ft}^2$

b. $35 \text{ yd} \cdot \dfrac{3 \text{ ft}}{1 \text{ yd}} = 105 \text{ ft}$

$\dfrac{128,100 \text{ ft}^2}{105 \text{ ft}} = 1220 \text{ ft}$

c. $2(105 \text{ ft}) + 2(1220 \text{ ft}) = 2650 \text{ ft}$

$2650 \text{ ft} \cdot \dfrac{1 \text{ yd}}{3 \text{ ft}} = 883\frac{1}{3} \text{ yd}$

12. a. $2(1.5 \text{ ft}) + 2(0.5\pi \text{ ft}) = (3.0 + \pi) \text{ ft} \approx$

6.14 ft

b. $10(13 \text{ m}) = 130 \text{ m}$

13. a. $S = 2(18 \text{ ft})(6 \text{ ft}) + 2(18 \text{ ft})(8 \text{ ft}) +$

$2(6 \text{ ft})(8 \text{ ft}) = 600 \text{ ft}^2$

$V = (18 \text{ ft})(6 \text{ ft})(8 \text{ ft}) = 864 \text{ ft}^3$

b. $S = 2\pi(6 \text{ in})(10 \text{ in}) \approx 602.88 \text{ in}^2$

$V = \pi(6 \text{ in})^2(10 \text{ in}) \approx 1130.4 \text{ in}^3$

Chapter 2 Test (con't)

14.

	Inequality	Interval	Line Graph
0 to 20	$0 \leq x \leq 20$	[0, 20]	
more than 20 & less than 50	$20 < x < 50$	(20, 50)	
50 or greater	$x \geq 50$	[50, +∞)	

15.

x	y	$x + y$	$x - y$	xy	$x \div y$
-8	4	-8 + 4 = -4	-8 - 4 = -12	(-8)(4) = -32	-8 ÷ 4 = -2
-6	-8 - (-6) = -2	-8	-6 - (-2) = -4	(-6)(-2) = 12	-6 ÷ (-2) = 3
6 *	-3	6 + (-3) = 3	9	-18	6 ÷ (-3) = -2
3 *	-6	3 + (-6) = -3	9	-18	$3 \div (-6) = -\frac{1}{2}$

* There are two possible combinations that satisfy x - y = 9 and (x)(y) = -18.

Cumulative Review

1. a.

x	$y = x^2 - 1$
-2	$(-2)^2 - 1 = 4 - 1 = 3$
-1	$(-1)^2 - 1 = 1 - 1 = 0$
0	$(0)^2 - 1 = 0 - 1 = -1$
1	$(1)^2 - 1 = 1 - 1 = 0$
2	$(2)^2 - 1 = 4 - 1 = 3$
3	$(3)^2 - 1 = 9 - 1 = 8$

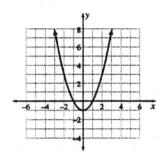

c.

x	$y = 2x + 3$
-2	$2(-2) + 3 = -4 + 3 = -1$
-1	$2(-1) + 3 = -2 + 3 = 1$
0	$2(0) + 3 = 0 + 3 = 3$
1	$2(1) + 3 = 2 + 3 = 5$
2	$2(2) + 3 = 4 + 3 = 7$
3	$2(3) + 3 = 6 + 3 = 9$

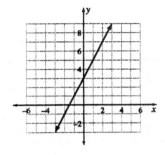

b.

x	$y = 2 - x$
-2	$2 - (-2) = 2 + 2 = 4$
-1	$2 - (-1) = 2 + 1 = 3$
0	$2 - (0) = 2$
1	$2 - (1) = 1$
2	$2 - (2) = 0$
3	$2 - (3) = -1$

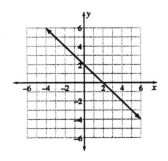

d.

x	$y = -2x$
-2	$-2(-2) = 4$
-1	$-2(-1) = 2$
0	$-2(0) = 0$
1	$-2(1) = -2$
2	$-2(2) = -4$
3	$-2(3) = -6$

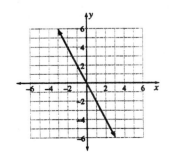

Cumulative Review (con't)

1. e.

x	$y = -x + 1$
-2	$-(-2) + 1 = 2 + 1 = 3$
-1	$-(-1) + 1 = 1 + 1 = 2$
0	$-(0) + 1 = 0 + 1 = 1$
1	$-(1) + 1 = 0$
2	$-(2) + 1 = -1$
3	$-(3) + 1 = -2$

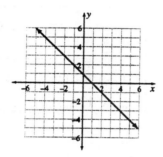

f.

x	$y = \left| x - 1 \right|$
-2	$\left| -2 - 1 \right| = \left| -3 \right| = 3$
-1	$\left| -1 - 1 \right| = \left| -2 \right| = 2$
0	$\left| 0 - 1 \right| = \left| -1 \right| = 1$
1	$\left| 1 - 1 \right| = \left| 0 \right| = 0$
2	$\left| 2 - 1 \right| = \left| 1 \right| = 1$
3	$\left| 3 - 1 \right| = \left| 2 \right| = 2$

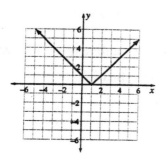

3. a. 4 is less than x.

b. The product of 3 and x is subtracted from 5.

c. Opposite of the opposite of x.

d. The absolute value of the difference between 3 and x.

5. a. $3(x + 1) + 5(x - 2) =$

$3x + 3 + 5x - 10 = 8x - 7$

b. $8 - 2(x - 1) = 8 - 2x + 2$

$= -2x + 10$

c. $8(x + 2) - 3(x - 2) =$

$8x + 16 - 3x + 6 = 5x + 22$

d. $5 - 3(2 - x) = 5 - 6 + 3x = 3x - 1$

Cumulative Review (con't)

7.

x	y	$x + y$	$x - y$	xy	$x \div y$
-4	$6 = 2 -(-4)$	2	$-4 - 6 = -10$	$(-4)(6) = -24$	$\frac{-4}{6} = -\frac{2}{3}$
5	$-3 = -15 \div 5$	$5 + (-3) = 2$	$5 - (-3) = 8$	-15	$\frac{5}{-3} = -\frac{5}{3}$
$-\frac{2}{3} = -\frac{8}{12}$	$\frac{3}{4} = \frac{9}{12}$	$-\frac{8}{12} + \frac{9}{12} = \frac{1}{12}$	$-\frac{8}{12} - \frac{9}{12} = -\frac{17}{12}$	$-\frac{2}{3} \cdot \frac{3}{4} = -\frac{1}{2}$	$-\frac{2}{3} \div \frac{3}{4} = -\frac{2}{3} \cdot \frac{4}{3}$ $= -\frac{8}{9}$
$\frac{7}{8} = \frac{35}{40}$	$-\frac{7}{10} = -\frac{28}{40}$	$\frac{35}{40} + (-\frac{28}{40}) = \frac{7}{40}$	$\frac{35}{40} - (-\frac{28}{40}) = \frac{63}{40}$	$\frac{7}{8} \cdot (-\frac{7}{10}) = -\frac{49}{80}$	$\frac{7}{8} \div (-\frac{7}{10}) =$ $\frac{7}{8} \cdot (-\frac{10}{7}) = -\frac{5}{4}$
1.44	1.8	$1.44 + 1.8 =$ 3.24	$1.44 - 1.8 =$ -0.36	$1.44(1.8) =$ 2.592	$1.44 \div 1.8 =$ 0.8
0.25	-0.5	$0.25 + (-0.5) =$ -0.25	$0.25 - (-0.5) =$ 0.75	$0.25(-0.5) =$ -0.125	$0.25 \div (-0.5) =$ -0.5

9. a. $+68 - 74 - 26 + 32 + 14 =$

$+68 + 32 - 74 - 26 + 14 =$

$+100 - 100 + 14 = 14$

b. $-16 + 18 - 35 + 12 - 15 - 24 =$

$-16 - 24 + 18 + 12 - 35 - 15 =$

$-40 + 30 - 50 = -60$

c. $5 \div \frac{2}{3} \cdot 4 = 5 \cdot \frac{3}{2} \cdot 4 = 5 \cdot 3 \cdot 2 = 30$

d. $7 \div \frac{3}{2} \cdot 6 = 7 \cdot \frac{2}{3} \cdot 6 = 7 \cdot 2 \cdot 2 = 28$

e. $\dfrac{a}{b} \div \dfrac{a}{b} = \dfrac{a}{b} \cdot \dfrac{b}{a} = 1$

f. $\dfrac{x}{y} \div \dfrac{-x}{y} = \dfrac{x}{y} \cdot \dfrac{y}{-x} = -1$

9. g. $\dfrac{6a + 2}{3} = \dfrac{6a}{3} + \dfrac{2}{3} = 2a + \frac{2}{3}$

h. $\dfrac{xy - x^2}{x} = \dfrac{xy}{x} - \dfrac{x^2}{x} = y - x$

11. $(40 \text{ in})(8 \text{ in})(20 \text{ ft}) \left(\dfrac{1 \text{ ft}}{12 \text{ in}}\right)^2 \left(\dfrac{1 \text{ yd}}{3 \text{ ft}}\right)^3 =$

$\dfrac{6400 \text{ in}^2\text{ft} \cdot 1 \text{ ft}^2 \cdot 1 \text{ yd}^3}{144 \text{ in}^2 \cdot 27 \text{ ft}^3} \approx 1.65 \text{ yd}^3$

13. a The distance from Erie to Pittsburgh equals the distance from Pittsburgh to Erie. (for example)

b. The commutative properties of addition and multiplication.

Section 3.1

1. $\dfrac{28}{35} = \dfrac{7 \cdot 4}{7 \cdot 5} = \dfrac{4}{5}$

3. $\dfrac{45}{6} = \dfrac{42+3}{6} = \dfrac{42}{6} + \dfrac{3}{6} = 7\frac{1}{2}$

5. $0.3 = \dfrac{30}{100} = 30\%$

7. $8 - 3(2 - 8) = 8 - 6 + 24 = 26$

9. **a.** conditional, 2 variables x, y

   b. conditional, 1 variable x

   c. conditional, 3 variables A, b, h

   d. identity

   e. identity

   f. identity

11. $\dfrac{x}{2} = 8$

13. $6 + x = 4$

15. $2x - 15 = -9$

17. $19 = 3 + 2x$

19. $3x - 4 = 17$

21. $26 - 4x = 2$

23. Four less than a number is six.

25. Five less than the product of three and a number is sixteen.

27. Two thirds of a number is twenty-four.

29. Six less than twice a number is ten.

31. **a.** $2(4) = 8$; Correct solution

   b. $(-6) + 1 = -5$; Correct solution

   c. $(12) - 3 = 15$; Incorrect - 3 was subtracted from 15, need to add 3.

33. **a.** $55(2) = 440$; Incorrect, 440 was divided by 22 not 55.

   b. $\frac{1}{2}(5) = 10$; Incorrect, 10 was divided by 2, need to multiply by 2.

   c. $1.05(40) = 42$; Correct solution

35. $x - 5 = 8$

   $ + 5 \quad + 5 \qquad$ Add 5

   $x = 13$

37. $9 = x + 12$

   $ - 12 \quad - 12 \qquad$ Subtract 12

   $x = -3$

39. $x - 6 = -10$

   $ + 6 \quad + 6 \qquad$ Add 6

   $x = -4$

41. $2x = 26$

   $\div 2 \quad \div 2 \qquad$ Divide by 2

   $x = 13$

Section 3.1 (con't)

43. $3 = 8x$

$\div 8 \quad \div 8$ Divide by 8

$x = \dfrac{3}{8}$

45. $\dfrac{x}{4} = 16$

$\cdot 4 \quad \cdot 4$ Multiply by 4

$x = 64$

47. $\dfrac{-x}{12} = 4$

$\cdot -12 \quad \cdot -12$ Multiply by -12

$x = -48$

49. $-4 = \frac{1}{2}x$

$\cdot 2 \quad \cdot 2$ Multiply by 2

$x = -8$

51. $-\frac{3}{4}x = 12$

$\cdot -\frac{4}{3} \quad \cdot -\frac{4}{3}$ Multiply by $-\frac{4}{3}$

$x = \dfrac{12 \cdot 4}{3} = 16$

53. f. x times -2 plus 4 gives 3

$4 - 2x = 3$

$-4 \qquad -4$

$-2x = 3 - 4 = -1$

$\div -2 \qquad \div -2$

$x = \frac{1}{2}$

55. c. x times $\frac{1}{2}$ subtract 4 gives 3.

$\frac{1}{2}x - 4 = 3$

$\quad + 4 \ + 4$

$\frac{1}{2}x = 7$

$\cdot 2 \quad \cdot 2$

$x = 14$

57. b. x times $\frac{1}{4}$ subtract 2 gives 3.

$\frac{1}{4}x - 2 = 3$

$\quad + 2 \ + 2$

$\frac{1}{4}x = 5$

$\cdot 4 \quad \cdot 4$

$x = 20$

59. $4x + 3 = 23$

$4x = 23 - 3, \ 4x = 20$

$x = 20 \div 4, \ x = 5$

61. $3x - 2 = 43$

$3x = 43 + 2, \ 3x = 45$

$x = 45 \div 3, \ x = 15$

Section 3.1 (con't)

63. $10 - x = -2$

$-x = -2 - 10, \ -x = -12$
$x = -12 \div -1, \ x = 12$

65. $3 = 10 - x$

$x + 3 = 10$
$x = 10 - 3, \ x = 7$

67. $0 = 3 - 3x$

$0 + 3x = 3, \ 3x = 3$
$x = 3 \div 3, \ x = 1$

69. $3 - 3x = 9$

$-3x = 9 - 3, \ -3x = 6$
$x = 6 \div -3, \ x = -2$

71. $\frac{1}{2}x + 3 = -2$

$\frac{1}{2}x = -2 - 3, \ \frac{1}{2}x = -5$
$x = -5 \cdot 2, \ x = -10$

73. $0 = \frac{3}{2}x + 3$

$\frac{3}{2}x = 0 - 3, \ \frac{3}{2}x = -3$
$x = -3 \cdot \frac{2}{3}, \ x = -2$

75. $-4 = \frac{2}{3}x - 2$

$-4 + 2 = \frac{2}{3}x, \ -2 = \frac{2}{3}x$
$-2 \cdot \frac{3}{2} = x, \ x = -3$

77. $4.2x - 3 = -9.3$

$4.2x = -9.3 + 3, \ 4.2x = -6.3$
$x = -6.3 \div 4.2, \ x = -1.5$

79. $7.5 = 4.2x - 3$

$4.2x = 7.5 + 3, \ 4.2x = 10.5$
$x = 10.5 \div 4.2, \ x = 2.5$

81. The first equation was divided by 5 on both sides.

83. 5 was subtracted from both sides of the first equation.

85. 4 was added to both sides of the first equation.

87. Both sides of the first equation were multiplied by 2.

89. $110 = 55t; \ t = \dfrac{110}{55}, \ t = 2$

91. $212 = \frac{9}{5}C + 32$

$\frac{9}{5}C = 212 - 32, \ \frac{9}{5}C = 180$
$C = 180 \cdot \frac{5}{9}, \ C = 100$

93. $200 = 55t; \ t = \dfrac{200}{55}, \ t = 3\frac{7}{11}$

95. $200 = 3r$

$r = \dfrac{200}{3}, \ r = 66\frac{2}{3}$

97. $\$0.10 = 0.06p$

$p = \dfrac{\$0.10}{0.06}, \ p \approx \$1.67$

Section 3.1 (con't)

99. $\$75 = \$5 + \$0.03715x$

$\$0.03715x = \$75 - \$5$
$x = \$70 \div \$0.03715,$
$x = 1884.25 \, \text{kwh}$

101. $15 = \frac{1}{2} \cdot 3h$

$h = 15 \cdot \frac{2}{3}, \quad h = 10$

103   Subtraction is the same as adding the opposite.

105.   Add a to each side, $x = b + a$.

107.   An identity equation is true for all values of the variable(s) while a conditional equation is true for only certain values.

Section 3.2

1. **a.** Linear equation.

   b. Linear equation.

   c. Non-linear, can not be written in $y = mx + b$ form.

   d. Linear equation.

   e. Non-linear, can not be written in $y = mx + b$ form.

   f. Linear equation.

3. $y = 2x + 5$

5. $y = 3x - 6$

7. $y = \frac{1}{2}x - 5$

9. The tip depends on the cost of the meal so y will represent the tip and x will be the cost of the meal.

   Remember to convert the % to a decimal.

   $y = 0.15x$

11. The Medicare payment depends on the wage so y is the payment and x represents wages.

    $y = 0.0145x$

13. The total cost depends on the number of pounds so y will represent the total cost and x will be the number of pounds.

    $y = \$1.49x$

15. The distance traveled depends on the speed so y will be distance and x will be speed.

    $y = 3x$

17. $y = \$75x + \$32$

19. $y = -\$3.25x + \$26$

21. **a.** From the table; when y = 8, x = 2

    b. From the table; when y = 6, x = 4

    c. Extending the table; when y = 3, x = 7

    d. Extending the table; when y = -2, x = 12

23. **a.** From the table; when y = -2, x = 1

    b. From the table; when y = -6, x = 2

    c. Extending the table; when y = 6, x = -1

    d. Extending the table; when y = -8, x = $2\frac{1}{2}$

Section 3.2 (con't)

25. $8 = 2x + 1$

$7 = 2x, \ x = 3\frac{1}{2}$

$10 = 2x + 1$
$9 = 2x, \ x = 4\frac{1}{2}$

$12 = 2x + 1$
$11 = 2x, \ x = 5\frac{1}{2}$

27. $12 = 5x - 4$

$16 = 5x, \ x = 3\frac{1}{5}$

$13 = 5x - 4$
$17 = 5x, \ x = 3\frac{2}{5}$

$14 = 5x - 4$
$18 = 5x, \ x = 3\frac{3}{5}$

$15 = 5x - 4$
$19 = 5x, \ x = 3\frac{4}{5}$

29. a. From the graph; when $y = 1$,
$x = -4$

b. From the graph, when $y = 0$,
$x = -6$

c. From the graph, when $y = -2$,
$x = -10$

31. a. From the graph; when $y = 1$,
$x = 1$

b. From the graph; when $y = -7$,
$x = -1$

31. c. From the graph; when $y = 5$,
$x = 2$

33. a. From the graph; when $y = -3$
there are 2 possible values for x,
$\{-2, 2\}$

b. From the graph; when $y = 3$ there
are 2 possible values for x, $\{-1, 1\}$

c. From the graph; when $y = 5$
there is only one value for x, $\{0\}$

d. $y = 6$ is not on the graph so there
are no values for x, $\{\}$

35. a. $3x - 4y = -12$

$-4y = -3x - 12$
Let $y = 0$, to find x - intercept
$0 = \frac{3}{4}x + 3, \ -3 = \frac{3}{4}x$
$-4 = x, \ $ x - intercept $= (-4, 0)$
$y = \frac{3}{4}x + 3,$
Let x = 0, to find y - intercept
$y = \frac{3}{4}(0) + 3, \ y = 3,$
y - intercept $= (0, 3)$

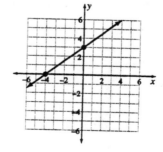

Section 3.2 (con't)

35. b. $2x + 5y = 10$

Let y = 0
$2x + 5(0) = 10$
$2x = 10, \ x = 5$
x - intercept $= (5, 0)$
Let x = 0
$2(0) + 5y = 10$
$5y = 10, \ y = 2$
y - intercept $= (0, 2)$

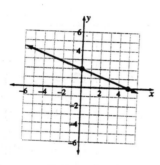

37. a. $y = \frac{2}{3}x - 4$

Let y = 0
$0 = \frac{2}{3}x - 4$
$4 = \frac{2}{3}x, \ x = 6$
x - intercept $= (6, 0)$

Let x = 0

y-intercept $= (0, -4)$

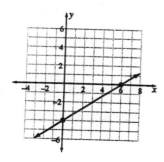

37. b. $y = \frac{3}{4}x + 3$

Let y = 0

$0 = \frac{3}{4}x + 3$
$-\frac{3}{4}x = 3, \ x = -4$
x - intercept $= (-4, 0)$

Let x = 0

y-intercept $= (0, 3)$

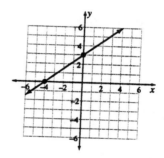

39. a. -2 = 3(0) - 2?

-2 = -2, true statement

b. 2 + 5(-2) = -8?

2 - 10 = -8, true statement

c. $4 = \frac{1}{2}(2)$?

4 = 1, false, (2, 4) is not a solution

d. 3 + 5 = 8?

8 = 8, true statement

e. $\frac{1}{2}(8)(9) = 36$?

4(9) = 36, true statement

Section 3.2 (con't)

41. A linear equation can be written in the form $y = mx + b$ or $mx + b = 0$.

43. Find n on the output side, locate the corresponding x value on the input side.

45. Find the value where the graph crosses the vertical axis; at $x = 0$.

47. {} is the set with nothing in it.

{0} contains the number zero.

49. True.

51. False

53. a. $3.50 = 5.00 - 0.05x$

$3.5 - 5.00 = -0.05x$

$-1.50 = -0.05x$

$\dfrac{-1.50}{-0.05} = x, \; x = 30$

b. $y = 5.00 - 0.05(0)$

$y = 5.00$

c. $0 = 5.00 - 0.05x$

$0.05x = 5.00$

$x = \dfrac{5.00}{0.05}, \; x = 1.00$

55. a. $2.25 = 5.00 - 0.05x$

$2.25 - 5.00 = -0.05x$

$\dfrac{-2.75}{-0.05} = x, \; x = 55 \text{ copies}$

b. equation c, $0 = 5.00 + 0.05x$

c. equation b, $y = 5.00 - 0.05(0)$

57. a.

x, min	y, $
0	12
10	12 - 0.30(10) = 9
20	12 - 0.30(20) = 6
30	12 - 0.30(30) = 3
40	12 - 0.30(40) = 0
50	0

b. $3.00

c. 10 minutes

d. $2 = 12 - 0.30x$

$-10 = -0.30x$

$\dfrac{-10}{-0.30} = x, \; x = 33\tfrac{1}{3} \text{ minutes}$

e. 40 minutes

f. Represents the total minutes used when the remaining value is zero.

g. $12

Section 3.2 (con't)

57. h. The initial value of the card before talking.

    i. $y = -\$0.30x + \$12$

59. a.

r, mph	D, miles
0	0
10	$3(10) = 30$
20	$3(20) = 60$
30	$3(30) = 90$
40	$3(40) = 120$

b. From the table, 15 miles is between 0 and 30 so her rate will be between 0 and 10 mph. Solving for rate: $15 = 3(r)$, $r = 5$ mph

c.   From the table, 75 miles is between 60 and 90 so his rate will be between 20 and 30 mph. Solving for rate: $75 = 3(r)$, $r = 25$ mph

d. (0, 0), The distance traveled at zero mph.

Mid-Chapter 3 Test

1. **a.** Conditional equation

   b. Identity

   c. Identity

   d. Conditional equation

2. **a.** Equivalent, 5 was subtracted from both sides of the first equation to obtain the second.

   b. Not equivalent, half of 8 is not 16

   c. Not equivalent, 2x + 3x is 5x

   d. Not equivalent, coefficient on x changed.

3. $3(-4) - 4 = -16$?

   $-12 - 4 = 16$, true statement

4. $4(-3) - 3 = -9$?

   $-12 - 3 = -9$, false statement

   $x = -3$ is not a solution

5. $-8(\frac{1}{2}) + 5 = 1$?

   $-4 + 5 = 1$, true statement

6. $-6 - 10(\frac{1}{2}) = -11$?

   $-6 - 5 = -11$, true statement

7. $x - 4 = 3$, $x = 3 + 4$, $x = 7$

8. $\frac{2}{3}x = 24$, $x = \dfrac{24 \cdot 3}{2}$, $x = 36$

9. $2x + 3 = -7$, $2x = -10$

   $x = \dfrac{-10}{2}$, $x = -5$

10. $\frac{1}{2}x - 8 = -1$, $\frac{1}{2}x = 7$

    $x = 7(2)$, $x = 14$

11. $3x = \frac{1}{2}$, $x = \frac{1}{2} \cdot \frac{1}{3}$, $x = \frac{1}{6}$

12. $3 - 2x = 8$, $-2x = 5$,

    $x = \dfrac{5}{-2}$, $x = -2\frac{1}{2}$

13. $2x + 5 = 10$

14. $5 = \frac{1}{2}x - 6$

15. $\$545 = \$85x + \$35$

16. $\$7.20 = -\$0.40x + \$20$

17. Five is four less than three times a number.

18. Four more than twice a number is -3.

19. **a.** From the table, when $y = 5$, $x = 2$

    b. From the table, when $y = 8$, $x = 8$

    c. Extending the table, when $y = 4$, $x = 0$

    d. Extending the table, when $y = 6.5$, $x = 5$

Mid-Chapter 3 Test (con't)

20. a. From the graph, when $y = -5$,

$x = 2$

b. From the graph, when $y = 1$,

$x = -4$

c. From the graph, when $y = -2$,

$x = -1$

21. a. From the graph, when $y = 6$,

$x = \{-4, 3\}$

b. From the graph, when $y = 0$,

$x = \{-3, 2\}$

c. From the graph, when $y = -6$,

$x = \{-1, 0\}$

22. There is no equal sign between 10
and x. A comma would be correct.

23. a.

x, children	y, $
5	70
10	70
15	70
20	$70 + 3(5) = 85$

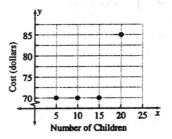

b. $70

c $85

d. $70, Cost for 0 children, not
meaningful.

e. $100 = $3(x - 15) + $70

Section 3.3

1.

x	$5x - 8$	$2(x + 2)$
-1	$5(-1) - 8 = -13$	$2(-1 + 2) = 2$
0	$5(0) - 8 = -8$	$2(0 + 2) = 4$
1	$5(1) - 8 = -3$	$2(1 + 2) = 6$
2	$5(2) - 8 = 2$	$2(2 + 2) = 8$
3	$5(3) - 8 = 7$	$2(3 + 2) = 10$
4	$\mathbf{5(4) - 8 = 12}$	$\mathbf{2(4 + 2) = 12}$

Common ordered pair $(4, 12)$, $x = 4$

3.

x	$3(x - 3)$	$6(x - 2)$
-1	$3(-1 - 3) = -12$	$6(-2 - 2) = -18$
0	$3(0 - 3) = -9$	$6(0 - 2) = -12$
1	$\mathbf{3(1 - 3) = -6}$	$\mathbf{6(1 - 2) = -6}$
2	$3(2 - 3) = -3$	$6(2 - 2) = 0$
3	$3(3 - 3) = 0$	$6(3 - 2) = 6$
4	$3(4 - 3) = 3$	$6(4 - 2) = 12$

Common ordered pair $(1, 6)$, $x = 1$

5. a. $3(2 - x) = 4 - x$ when $x = 1$

(intersection of 2 lines)

b. $3(2 - x) = 6$ when $x = 0$

(y-intercept)

c. $4 - x = 6$ when $x = -2$

d. $4 - x = 2$ when $x = 2$

e. $3(2 - x) = 0$ when $x = 2$

(x-intercept)

7. a. $3(3 + x) = 2(2 - x)$ when $x = -1$

(intersection of 2 lines)

b. $3(3 + x) = 0$ when $x = -3$

(x-intercept)

c. $2(2 - x) = 4$ when $x = 2$

(x-intercept)

d. $2(2 - x) = 4$ when $x = 0$

(y-intercept)

e. $3(3 + x) = 3$ when $x = -2$

Section 3.3 (con't)

9. a.

x	$y = 8 - 3(x + 2)$
-3	$8 - 3(-3 + 2) = 11$
-2	$8 - 3(-2 + 2) = 8$
-1	$8 - 3(-1 + 2) = 5$
0	$8 - 3(0 + 2) = 2$
1	$8 - 3(1 + 2) = -1$
2	$8 - 3(2 + 2) = -4$

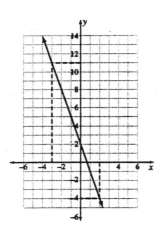

b. $8 - 3(x + 2) = 11$

$8 - 3x - 6 = 11, \ 2 - 3x = 11$

$-3x = 9, \ x = -3$

c. $8 - 3(x + 2) = 0$

$8 - 3x - 6 = 0$

$2 - 3x = 0$

$-3x = -2, \ x = \frac{2}{3}$

9. d. $8 - 3(x + 2) = -4$

$8 - 3x - 6 = -4$

$2 - 3x = -4$

$-3x = -6, \ x = 2$

11. $2(x - 3) = 0$

$2x - 6 = 0, \ 2x = 6$

$x = 3$

13. $-2(x + 1) = 5$

$-2x - 2 = 5, \ -2x = 7$

$x = -\frac{7}{2}, \ x = -3\frac{1}{2}$

15. $2(3x + 1) = 5 - 3x$

$6x + 2 = 5 - 3x$

$9x = 3, \ x = \frac{1}{3}$

17. $4x + 6 = 2(1 + 3x) + 1$

$4x + 6 = 2 + 6x + 1 = 3 + 6x$

$3 = 2x, \ x = \frac{3}{2}, \ x = 1\frac{1}{2}$

19. $7(x + 1) = 11 + x$

$7x + 7 = 11 + x$

$6x = 4, \ x = \frac{2}{3}$

21. $2(x + 3) = 3x - 2$

$2x + 6 = 3x - 2$

$x = 8$

Section 3.3 (con't)

23. $\frac{3}{5}x = 15$

$x = \dfrac{15 \cdot 5}{3}, \; x = 25 \;\checkmark$

25. $\frac{1}{3}x = x - 12$

$-\frac{2}{3}x = -12, \; x = 18$

27. $\frac{3}{4}x = x + 3$

$-\frac{1}{4}x = 3, \; x = -12$

29. $\frac{1}{2}(x + 5) = x - 4$

$x + 5 = 2x - 8$

$x = 13$

31. $6 - 4(x - 2) = 22$

$6 - 4x + 8 = 22, \; 14 - 4x = 22$

$-4x = 8, \; x = -2$

33. $2(8 - x) = 1 + 4x$

$16 - 2x = 1 + 4x$

$15 = 6x, \; x = \frac{15}{6}, \; x = 2\frac{1}{2}$

35. $2 - 5(x + 1) = 4 - 3x$

$2 - 5x - 5 = 4 - 3x,$

$-5x - 3 = 4 - 3x$

$-2x = 7, \; x = -\frac{7}{2}, \; x = -3\frac{1}{2}$

37. $7 - 2(x - 1) = 5 - 3x$

$7 - 2x + 2 = 5 - 3x$

$9 - 2x = 5 - 3x$

$x = -4$

39. $5 - 2(x - 3) = 3(x - 3)$

$5 - 2x + 6 = 3x - 9$

$11 - 2x = 3x - 9$

$11 + 9 = 3x + 2x$

$20 = 5x, \; x = 4$

41. Distribute the 2; Distributive property

43. Add -7 to both sides; Addition property

45. Add x to both sides; Addition property

47. Let x be a number; $2(x + 5) = 14$

$2x + 10 = 14, \; 2x = 4, \; x = 2$

49. Let x be a number; $-2(x - 4) = 6$

$x - 4 = -3, \; x = 1$

51. Let x = input, y = output;

$y = \frac{1}{2}(x - 5)$

53. Let x = input, y = output;

$y = 15(3 - x)$

Section 3.3 (con't)

55. Let y be the handicap and x be the bowling average;

$y = 0.80(200 - x)$

57. $100 = \$0.15(x - 100) + \$65$

$\$35 = \$0.15x - \$15$

$\$50 = \$0.15x, \ x \approx 333$ miles

59. $\$6.50 = \$1.50(x - 1) + \$2$

$\$4.50 = \$1.50x - \$1.50$

$\$6.00 = \$1.50x, \ x = 4$ copies

61.

	4	5	6
Let $x = 4$	x	$x + 1$	$x + 2$
Let $x = 5$	$x - 1$	x	$x + 1$
Let $x = 6$	$x - 2$	$x - 1$	x

63.

	5	7	9
Let $x = 5$	x	$x + 2$	$x + 4$
Let $x = 7$	$x - 2$	x	$x + 2$
Let $x = 9$	$x - 4$	$x - 2$	x

65. $x + (x + 1) + (x + 2) = 42$

$3x + 3 = 42, \ 3x = 39, x = 13$

The integers are 13, 14, 15.

67. $x + (x + 2) + (x + 4) = 177$

$3x + 6 = 177, \ 3x = 171, \ x = 57$

The integers are 57, 59, 61.

69.

There are 2 units between consecutive odd numbers and 2 units between consecutive even number.

71. 2012 is divisible by 4,

$x, x + 4, x + 8, x + 12$

73. $x + (x + 7) + (x + 14) + (x + 21) +$
$(x + 28) + (x + 35) + (x + 42) =$
$13{,}741, \ 7x + 147 = 13{,}741,$
$7x = 13{,}594, \ x = 1942$

75. $40 = 0.10x, \ x = \dfrac{40}{0.10}, \ x = 400$

77. $35 = 0.10x, \ x = \dfrac{35}{0.10}, \ x = 350$

79. $550 = 50 + (x - 500)$

$500 = x - 500, \ x = 1000$

81. Find the common ordered pairs in tables built from each side of the equation.

Section 3.3 (con't)

83. Add or subtract to move variable terms to one side and constant terms to the other side. Add like terms. Divide by the numerical coefficient on x.

85. Subtract c from both sides. Multiply by $\frac{b}{a}$.

87. $x = 1, 1 + 1 = 2, 1 + 3 = 4;$

$x = 2, 2 + 1 = 3, 2 + 3 = 5$

89. x is a positive number.

Section 3.4

1. $(5, +\infty)$

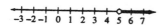

3. $(-\infty, -2]$

5. $[0, +\infty)$

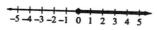

7. $(-\infty, -1)$

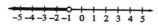

9. $x > 2, (2, +\infty)$

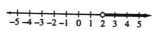

11. $x < 1, (-\infty, 1)$

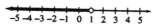

13. $x > 2, (2, +\infty)$

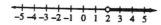

15. $x < 4, (-\infty, 4)$

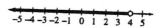

17. $x > -2, (-2, +\infty)$

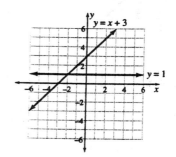

19. $x \geq 4, [4, +\infty)$

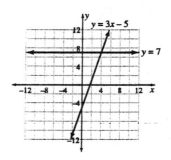

Section 3.4 (con't)

21. $x < 1$, $(-\infty, 1)$

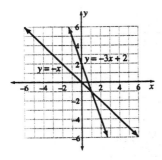

23. $x < 1$, $(-\infty, 1)$

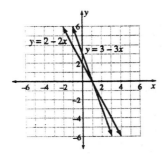

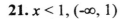

25. $-3 > -3x + 3$

$-6 > -3x$, $2 < x$

$x > 2$, $(2, +\infty)$

27. $0 < -3x + 3$

$3x < 3$, $x < 1$, $(-\infty, 1)$

29. $-x + 4 < \frac{1}{2}x + 1$

$3 < \frac{3}{2}x$, $2 < x$

$x > 2$, $(2, +\infty)$

31. $-x + 4 > 0$, $4 > x$

$x < 4$, $(-\infty, 4)$

33. $-1 < -3x + 2$, $3x < 3$

$x < 1$, $(-\infty, 1)$

35. $-2 \le 2 - 2x$, $2x \le 4$

$x \le 2$, $(-\infty, 2]$

37. $3 - 2x \le -5$, $8 \le 2x$, $4 \le x$

$x \ge 4$, $[4, +\infty)$

39. $-4 > x + 5$, $-9 > x$

$x < -9$, $(-\infty, -9)$

41. $1 - 4x < -4$, $5 < 4x$, $\frac{5}{4} < x$

$x > 1\frac{1}{4}$, $(1\frac{1}{4}, +\infty)$

43. $-2 > 2x + 1$, $-3 > 2x$, $-\frac{3}{2} > x$

$x < -1\frac{1}{2}$, $(-\infty, -1\frac{1}{2})$

45. $2x < x + 5$, $x < 5$, $(-\infty, 5)$

47. $4 - 2x \ge x - 2$, $6 \ge 3x$, $2 \ge x$

$x \le 2$, $(-\infty, 2]$

Section 3.4 (con't)

49. $3x - 4 < -2x + 1$, $5x < 5$,

$x < 1$, $(-\infty, 1)$

51. $2(x + 3) < 5x$, $2x + 6 < 5x$

$6 < 3x$, $2 < x$

$x > 2$, $(2, +\infty)$

53. $\frac{1}{2}x > 4 + x$, $x > 8 + 2x$

$-x > 8$, $x < -8$, $(-\infty, -8)$

55. $-x > -\frac{1}{2}x + 1$, $-\frac{1}{2}x > 1$

$x < -2$, $(-\infty, -2)$

57. Let x be the points needed on the final.

$$\frac{78 + 84 + 72 + 5 + x}{520} \geq 0.80$$

$239 + x \geq 416$, $x \geq 177$,

Student needs more points than the final exam is worth. Student can not earn a B.

59. Let x be the total points on the 3 earlier tests.

$$\frac{x + 70 + 135}{520} \geq 0.90$$

$x + 205 \geq 468$, $x \geq 263$ points

61. Let x be the final exam score.

$$\frac{88 + 84 + 89 + 70 + x}{100 + 100 + 100 + 70 + 200} \geq 0.90$$

$$\frac{331 + x}{570} \geq 0.90, \quad 331 + x \geq 513$$

$x \geq 182$ points

63. $\$5.50x + \$17 \leq \$160$

$\$5.50x \leq \$143$, $x \leq 26$ children

65. $\$8.50x + \$350 \leq \$1030$

$\$8.50x \leq \$680$, $x \leq 80$ people

67. $\$8.50x + \$350 \leq \$17.50x + \$100$

69. If $a < b$ then $\frac{a}{c} > \frac{b}{c}$ where a and b are real numbers and $c < 0$.

71. The inequality sign reverses,.

$+4 > +3$

Chapter 3 Review

1. Equivalent equations.

3. 2 variable equation and a conditional equation

5. Conditional equation

7. $-5(-1) + 4 = 9$?

   $5 + 4 = 9$, true statement

9. $8 - 7.5(0.4) = 5$?

   $8 - 3 = 5$, true statement

11. $x + 3 = -4$, $x = -7$

13. $3x = 27$, $x = 9$

15. $\frac{1}{2}x = 12$, $x = 24$

17. $4x - 2 = 22$, $4x = 24$, $x = 6$

19. $-2x + 3 = 9$, $-2x = 6$, $x = -3$

21. $-2(x - 4) = 18$, $x - 4 = -9$, $x = -5$

23. $5 - 2(x + 1) = -11$

   $5 - 2x - 2 = 3 - 2x = -11$

   $-2x = -14$, $x = 7$

25. $3x - 1 = x + 1$, $2x = 2$, $x = 1$

27. $-2(x - 3) = \frac{1}{2}x + 3$

   $-2x + 6 = \frac{1}{2}x + 3$, $3 = \frac{5}{2}x$

   $x = \frac{6}{5}$, $x = 1\frac{1}{5}$

29. $3x - 6 = -15$

31. $-7x = 21$

33. $3x = 2x + 4$

35. $6(2 + x) = -6$

37. The quotient of a number and 5 is 15.

39. Three times the difference between a number and four gives -18.

41. $12 - 1 = 11$, $12 + 1 = 13$

43. $15 + 3 = 18$, $15 + 6 = 21$

45. $9 = 10 - 1 = x - 1$

   $11 = 10 + 1 = x + 1$

47. $-5 = -3 - 2 = x - 2$

   $-1 = -3 + 2 = x + 2$

49. $x + 2(x + 1) = 17$

   $x + 2x + 2 = 17$, $3x + 2 = 17$

   $3x = 15$, $x = 5$

51. **a** From the table, when $y = 4$, $x = 2$

   b. From the table, when $y = 13$, $x = 5$

   c. From the table, when $y = 8$, x is between 3 and 4. Solving for x:

   $8 = 3x - 2$; $10 = 3x$, $x = \frac{10}{3}$, $x = 3\frac{1}{3}$

   d. Extending the table, when $y = 16$, $x = 6$.

Chapter 3 Review (con't)

53. a. From the graph, when $y = 8$,

$x = \{2\}$

b. From the graph, when $y = 6$,

$x = \{1, 3\}$

c. From the graph, when $y = 0$,

$x = \{0, 4\}$

55. a. From the graph:

$3x - 3 = x + 1$ when $x = 2$

$3(2) - 3 = 2 + 1$?

$6 - 3 = 3$, checks

b. From the graph:

$x + 1 = -2$ when $x = -3$

$-3 + 1 = 2$, checks

c. From the graph:

$3x - 3 = -6$ when $x = -1$

$3(-1) - 3 = -6$?

$-3 - 3 = -6$, checks

57. Let y be distance and x be speed.

$y = 4x$, x and y intercepts at (0, 0),
0 distance traveled at 0 speed.

59. Let y be the tax and x be the cost of
the meal.

$y = 0.075x$, x and y intercepts at
(0, 0), 0 tax for 0 meal cost.

61. Let y be the total cost and x be the
number of credits.

$y = \$300x + \$150$,

$0 = \$300x + \$150$, $\$300x = -\$150$

$x = -0.5$, x-intercept at (-0.5, 0) is not

meaningful, y-intercept at (0, \$150)

is cost of fees.

63. Let y be the remaining value and x be

the number of visits.

$y = -\$10x + \$520$,

$0 = -\$10x + \$520$, $\$10x = \$520$,

$x = 52$, x-intercept at (52, 0) is the

total number of visits when

remaining value is zero.

y-intercept at   (0, \$520) is value

with zero visits.

65. $440 = 20(x + 3)$

$440 = 20x + 60$

$380 = 20x$, $19 = x$

19 seats in original row

67. $x + (x + 5) + (x + 10) = 60$

$3x + 15 = 60$, $3x = 45$

$x = 15, x + 5 = 20, x + 10 = 25$

69. a, b, c does not indicate that the

numbers are 1 unit apart.

Chapter 3 Review (con't)

71. Student has earned 244 of a possible 370 points so far.

$$\frac{244 + x}{370 + 150} \geq 0.80, \ \frac{244 + x}{520} \geq 0.80$$

$244 + x \geq 416, \ x \geq 172$ points, not possible.

73. $2 < x - 3, \ 5 < x$, Solution a, $x > 5$

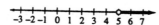

75. $1 < x - 4, \ 5 < x$

$x > 5, (5, +\infty)$

77. $-\frac{1}{2}x > 8, \ x < -16, (-\infty, -16)$

79. $5 - 3x > 13, \ -3x > 8, \ x < -\frac{8}{3}$

$x < -2\frac{2}{3}, (-\infty, -2\frac{2}{3})$

81. $13 \leq 7 - \frac{x}{3}, 6 \leq -\frac{x}{3}, -18 \geq x$

$x \leq -18, (-\infty, -18]$

83. $15 - 2x < x - 6$

$21 < 3x, 7 < x,$

$x > 7, (7, +\infty)$

85.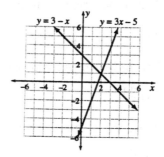

$3x - 5 \leq 3 - x, 4x \leq 8$

$x \leq 2, (-\infty, 2]$

87. $-2x < 1 - x, -x < 1,$

$x > -1, (-1, +\infty)$

89. $x > 2x + 3, -3 > x,$

$x < -3, (-\infty, -3)$

91. $x + 1 \leq 3 - x, 2x \leq 2$

$x \leq 1, (-\infty, 1]$

93. $-2x + 2 > -2 - x, 4 > x,$

$x < 4, (-\infty, 4)$

95. $3x + 2 \geq 3 - 2x, 5x \geq 1$

$x \geq \frac{1}{5}, [\frac{1}{5}, +\infty)$

97. $\$25x + \$550 \leq \$3500$

$\$25x \leq \$2950, x \leq 118$ people

99. $\$12x + \$250 < \$8x + \$320$

$\$4x < \$70, \ x < 17.5$ feet

Chapter 3 Test

1. $x + 8 = -3$, $x = -11$

2. $4 - x = 5$, $-x = 1$, $x = -1$

3. $\frac{2}{5}x = 30$, $x = \dfrac{30 \cdot 5}{2}$, $x = 75$

4. $-6x = 3$, $x = \frac{3}{-6}$, $x = -\frac{1}{2}$

5. $5 - 2x = 3$, $2 = 2x$, $x = 1$

6. $\frac{1}{2}x + 5 = -3$, $\frac{1}{2}x = -8$, $x = -16$

7. $2(x - 2) = -x - 1$, $2x - 4 = -x - 1$

   $3x = 3$, $x = 1$

8. $4(x - 3) = 6$, $4x - 12 = 6$

   $4x = 18$, $x = 4\frac{1}{2}$

9. $4 - 2(x - 4) = 2x$, $4 - 2x + 8 = 2x$

   $12 - 2x = 2x$, $12 = 4x$, $x = 3$

10. $-2(x - 3) = -0.5(x - 6)$

    $-2x + 6 = -0.5x + 3$, $3 = 1.5x$

    $x = 2$

11. solution (1, 3)

12. Zero is between -1 and 1 so x is between 2 and 3. The line is straight so because zero is half-way between -1 and 1 the value of x will be half-way between 2 and 3.

13. Find the point where the graph crosses the x-axis.

14. **a.** (1, -2)

    b. $-2 = 2(1) - 4$?

    $-2 = 2 - 4$, checks

    $-2 = -(1) - 1$?

    $-1 = -1 - 1$, checks

    c. $2x - 4 = -x - 1$, $3x = 3$, $x = 1$

    d. It is the x value of the point of intersection.

15. Let x be a number:

    $\frac{1}{2}x + 6 = 15$, $\frac{1}{2}x = 9$, $x = 18$

16. Let x be a number:

    $3x - 7 = -31$, $3x = -24$, $x = -8$

17. Let x be input, y be output:

    $y = \frac{1}{3}x$

18. Let x be input, y be output:

    $y = 2x - 2$

19. Let x be the first integer:

    $x + (x + 1) + (x + 2) + (x + 3) = -74$

    $4x + 6 = -74$, $4x = -80$, $x = -20$

Chapter 3 Test (con't)

20. $2x + 5 < -3$, $2x < -8$, $x < -4$, $(-\infty, -4)$

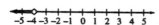

21. $3 - 2x > 11$, $-8 > 2x$, $-4 > x$

  $x < -4$, $(-\infty, -4)$

22. $2x + 8 \geq \frac{1}{2}(x + 1)$

  $4x + 16 \geq x + 1$, $3x \geq -15$

  $x \geq -5$, $[-5, +\infty)$

23. The x value of the intersection gives the end point for the solution.

  $15 - 2x > x - 6$, $21 > 3x$, $7 > x$,

  $x < 7$

24. a.

x	$y = 0.615x + 4$
0	$0.615(0) + 4 = 4$
20	$0.615(20) + 4 = 16.30$
40	$0.615(40) + 4 = 28.60$
60	$0.615(60) + 4 = 40.90$
80	$0.615(80) + 4 = 53.20$
100	$0.615(100) + 4 = 65.50$

b.

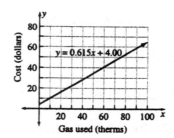

c. $y = \$0.615x + \$4$

d. $\$47.05 = \$0.615x + \$4$

  $\$43.05 = \$0.615x$

  $x = 70$ therms

e. $(-6.50, 0)$, no meaning

f. $(0, 4)$, monthly fee

Section 4.1

1. Interest in terms of principal, rate and time.

3. Rate in terms of distance and time.

5. Grade percent in terms of tests 1, 2, & 3, homework, final exam, and total points.

7. $A = \ell w$

9. $A = \pi r^2$

11. $A = \frac{1}{2} bh$

13. $P = 2\ell + 2w$

15. $C = fm$

Note to Students: In the following solutions the variable being solved for is placed on the left side of the equal sign in the final equation.

17. $p = 5n$,

$$\frac{p}{5} = \frac{5n}{5}$$

$$n = \frac{p}{5}$$

19. $A - P = H$,

$A - P + P = H + P$,

$A = H + P$

21. $C = 2\pi r$,

$$\frac{C}{2\pi} = \frac{2\pi r}{2\pi},$$

$$r = \frac{C}{2\pi}$$

23. $A = bh$,

$$\frac{A}{b} = \frac{bh}{b},$$

$$h = \frac{A}{b}$$

25. $I = prt$,

$$\frac{I}{pr} = \frac{prt}{pr},$$

$$t = \frac{I}{pr}$$

27. $C = \pi d$,

$$\frac{C}{\pi} = \frac{\pi d}{\pi},$$

$$d = \frac{C}{\pi}$$

29. $P = R - C$,

$P + C = R$,

$P + C - P = R - P$,

$C = R - P$

31. $PV = nRT$,

$$\frac{PV}{RT} = \frac{nRT}{RT},$$

$$n = \frac{PV}{RT}$$

Section 4.1 (con't)

33. $C_1V_1 = C_2V_2$,

$$\frac{C_1V_1}{C_1} = \frac{C_2V_2}{C_1},$$

$$V_1 = \frac{C_2V_2}{C_1}$$

35. $P = a + b + c$,

$$P - a - b = a + b + c - a - b,$$

$$c = P - a - b$$

37. $A = \frac{1}{2}h(a + b)$,

$$2A = 2 \cdot \frac{1}{2}h(a + b)$$

$$2A = h(a + b)$$

$$\frac{2A}{a + b} = \frac{h(a + b)}{a + b}$$

$$h = \frac{2A}{a + b}$$

39. $V = \frac{1}{3}\pi r^2 h$,

$$3V = 3 \cdot \frac{1}{3}\pi r^2 h,$$

$$3V = \pi r^2 h,$$

$$\frac{3V}{\pi h} = \frac{\pi r^2 h}{\pi h}$$

$$r^2 = \frac{3v}{\pi h}$$

41. $x = \dfrac{-b}{2a}$,

$$2ax = 2a\frac{-b}{2a},$$

$$2ax = -b,$$

$$(-1)2ax = (-1)(-b)$$

$$b = -2ax$$

43. $y = mx + b$,

$$y - mx = mx + b - mx$$

$$b = y - mx$$

45. $d^2 = \dfrac{3h}{2}$,

$$\frac{2d^2}{3} = \frac{2}{3} \cdot \frac{3h}{2},$$

$$h = \frac{2d^2}{3}$$

47. $t^2 = \dfrac{2d}{g}$,

$$gt^2 = g \cdot \frac{2d}{g},$$

$$gt^2 = 2d,$$

$$\frac{gt^2}{t^2} = \frac{2d}{t^2},$$

$$g = \frac{2d}{t^2}$$

49. $xy = -4$,

$$\frac{xy}{x} = \frac{-4}{x},$$

$$y = \frac{-4}{x}$$

Section 4.1 (con't)

51. $3x - y = 10,$

$3x - y + y = 10 + y,$
$3x = 10 + y,$
$3x - 10 = 10 + y - 10,$
$y = 3x - 10$

53. $x - 2y = -5,$

$x - 2y + 2y = -5 + 2y,$
$x = -5 + 2y,$
$x + 5 = -5 + 2y + 5,$
$x + 5 = 2y,$
$\frac{1}{2}(x + 5) = \frac{1}{2} \cdot 2y,$
$y = \frac{1}{2}x + \frac{5}{2}$

55. $3x - 2y = 6,$

$3x = 6 + 2y,$
$3x - 6 = 2y,$
$\frac{3x - 6}{2} = \frac{2y}{2},$
$y = \frac{3}{2}x - 3$

57. a. $A = P + Prt,$

$A - P = Prt$

$r = \dfrac{A - P}{Pt}$

b. $A = \$11,050, \ P = \$10,000$

$t = 2 \text{ years}$

$r = \dfrac{11,050 - 10,000}{10,000 \cdot 2} = \dfrac{1,050}{20,000}$

$r = 0.0525 \text{ or } 5\frac{1}{4}\%$

59. a. $C = \frac{5}{9}(F - 32),$

$\frac{9}{5}C = F - 32$

$F = \frac{9}{5}C + 32$

b. $C = 37°,$

$F = \frac{9}{5} \cdot 37 + 32, \ F = 98.6°$

61. a. $H = 0.8(200 - 140),$

$H = 0.8(60),$

$H = 48$

b. $H = 0.8(200 - A)$

$200 - A = \dfrac{H}{0.8},$

$A = 200 - \dfrac{H}{0.8}$

c. $H = 24,$

$A = 200 - \dfrac{24}{0.8},$
$A = 170$

63. a. $y = mx + b,$

$b = y - mx$

b. $b = 4 - 2(3), \ b = -2$

c. $b = 4 - (-2)(3), \ b = 10$

d. $b = 4 - \frac{1}{2}(3), \ b = 2\frac{1}{2}$

e. $b = 4 - (-\frac{1}{2})(3), \ b = 5\frac{1}{2}$

Section 4.2

1. **a.** There is exactly one output for each input, it is a function; inputs are the set of first numbers, {-6, -5, 5, 6}; outputs are the set of second numbers {5, 6}.

    b. Not a function; the input 5 has 2 different outputs, the input 6 has 2 different outputs.

3. **a.** Function, inputs {2, 3, 4} output $\{\frac{1}{2}, \frac{1}{3}, \frac{1}{4}\}$

    b. Not a function

5. Function, input {Eden, Tuckman, McClintock}, output {Barbara}

7. Not a function

9. Long distance is billed at cost per minute. input {units of time}, output {cost of call}

11. The tilt of the Earth gives long December days in the southern hemisphere and short days in the northern. input {distance from equator}, output {hours of sunlight}

13 - 17 Answers will vary

19. C = auto registration cost, v = value of auto, C = f(v)

21. C = circumference, d = diameter, C = f(d)

23.

x	$f(x) = 2x - 1$
-2	2(-2) - 1 = -5
-1	2(-1) - 1 = -3
0	2(0) - 1 = -1
1	2(1) - 1 = 1
2	2(2) - 1 = 3

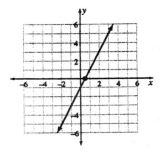

25.

x	$f(x) = 2 - 3x$
-2	2 - 3(-2) = 8
-1	2 - 3(-1) = 5
0	2 - 3(0) = 2
1	2 - 3(1) = -1
2	2 - 3(2) = -4

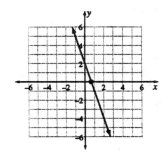

Section 4.2 (con't)

27.

x	$f(x) = \frac{1}{4}x + 1$
-2	$\frac{1}{4}(-2) + 1 = \frac{1}{2}$
-1	$\frac{1}{4}(-1) + 1 = \frac{3}{4}$
0	$\frac{1}{4}(0) + 1 = 1$
1	$\frac{1}{4}(1) + 1 = 1\frac{1}{4}$
2	$\frac{1}{4}(2) + 1 = 1\frac{1}{2}$

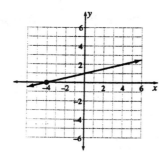

29. $f(x) = 0$,

$2x - 1 = 0$,

$2x = 1$,

$x = \frac{1}{2}$

31. $f(x) = 0$,

$2 - 3x = 0$,

$2 = 3x$,

$x = \frac{2}{3}$

33. $f(x) = 0$,

$\frac{1}{4}x + 1 = 0$,

$\frac{1}{4}x = -1$,

$x = -4$

Answers 29, 31 and 33 could have been found from the x-intercept on the graphs.

35. Function of x. One value of y for each x.

37. Function of x. One value of y for each x.

39. Function of x. One value of y for each x.

41. Not a function of x. Two possible values of y for each x.

43. Not a function of x. Infinite number of values of y for each x.

45. $h(4) = \left|\frac{1}{2}(4)\right| = |2| = 2$

$h(-4) = \left|\frac{1}{2}(-4)\right| = |-2| = 2$

47. $H(4) = 4 - 4^2 = 4 - 16 = -12$

$H(-4) = (-4) - (-4)^2 = -4 - 16 = -20$

49. $g(-2) = (-2)^2 + 1 = 4 + 1 = 5$

$g(1) = (1)^2 + 1 = 1 + 1 = 2$

51. $G(-2) = 2 - (-2)^2 = 2 - 4 = -2$

$G(1) = 2 - (1)^2 = 2 - 1 = 1$

Section 4.2 (con't)

53. (45) $h(0) = \left|\frac{1}{2}(0)\right| = 0$

(47) $H(0) = 0 - 0^2 = 0$

(49) $g(0) = 0^2 + 1 = 1$

(51) $G(0) = 2 - 0^2 = 2$

55. $f(x) = 0$

$3(x - 4) + x = 0,$

$3x - 12 + x = 0,$

$4x - 12 = 0$

$4x = 12,$

$x = 3$

57. $f(x) = 0,$

$9 - 3(x - 1) = 0,$

$9 - 3x + 3 = 0,$

$12 - 3x = 0,$

$12 = 3x$

$x = 4$

59. $f(x) = 0,$

$2x + 3(x - 5) = 0,$

$2x + 3x - 15 = 0,$

$5x - 15 = 0,$

$5x = 15,$

$x = 3$

61. $f(x) = 0,$

$115x + 48 = 0$

$115x = -48,$

$x = \dfrac{-48}{115}$

This value has no meaning.

$f(0) = 115(0) + 48 = 48$

This value is the fees included in the tuition.

63. $f(x) = 0,$

$19.50 - 0.75x = 0,$

$19.50 = 0.75x,$

$x = 26$

This is the total number of minutes the phone card will buy.

$f(0) = 19.50 - 0.75(0) = 19.50$

This is the original value of the phone card.

65. Not a function, there are 2 values of y for some x's.

67. Function

69. Not a function

71. Substitute a for every x in the equation defined by $f(x)$.

73. Find $f(0)$.

Section 4.2 (con't)

75. Answers will vary, multiplication is generally written $f \cdot x$ or fx

77. d = distance, r = rate, t = time

d = f(r, t)

79. V = volume, r = radius, h = height

V = f(r, h)

Mid-Chapter 4 Test

1. Temperature in Celsius in terms of temperature in Fahrenheit.

2. Volume of a sphere in terms of radius.

3. $3 = 4(-2) + b$,

   $3 = -8 + b$,

   $b = 11$

4. $-4 = \frac{2}{3}(-6) + b$,

   $-4 = -4 + b$,

   $b = 0$

5. $C = K - 273$,

   $K = C + 273$

6. $d^2 = \dfrac{3h}{8}$,

   $\dfrac{8}{3} \cdot d^2 = \dfrac{8}{3} \cdot \dfrac{3h}{8}$,

   $h = \dfrac{8d^2}{3}$

7. $l = a + (n - 1)d$,

   $l - a = (n - 1)d$,

   $d = \dfrac{l - a}{n - 1}$

8. **a.** $f(6) = 3(6) = 18$

   b. $f(6) = \frac{1}{2}(6) + 2 = 3 + 2 = 5$

   c. $f(6) = \frac{1}{2}(6 + 2) = \frac{1}{2}(8) = 4$

   d. $f(6) = (6)^2 - 2 = 36 - 2 = 34$

9.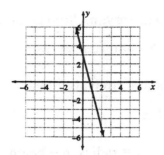

10. **a.** Yes, one value of y for each x.

    b. $\{2, 3, 4, 5, 6, 7\}$

    c. $\{3, 4\}$

11.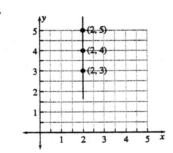

    a. one

    b. one input, three outputs

12. $f(0) = 3(0) - 5 = -5$

13. $f(x) = 0$,.

    $\frac{1}{2}x + 2 = 0$,

    $\frac{1}{2}x = -2$,

    $x = -4$

14. Vertical axis intercept

Section 4.3

1. **a.** Line rises from left to right, positive slope. Rise = 3, run = 1, slope = $\frac{3}{1} = 3$

   b. Line rises from left to right, positive slope. Rise = 2, run = 3, slope = $\frac{2}{3}$

   c. Horizontal line, zero slope. Rise = 0, slope = 0

3. **a.** Line descends from left to right, negative slope. Rise = -2, run = 1, slope = $\frac{-2}{1} = -2$

   b. Line rises from left to right, positive slope. Rise = 3, run = 2, slope = $\frac{3}{2}$

   c. Vertical line, undefined slope. Rise = 4, run = 0, slope = $\frac{4}{0}$, undefined

5. **a.** Line descends from left to right, negative slope. Rise = -50, run = 1, slope = $\frac{-50}{1} = -50$

   b. Line rises from left to right, positive slope. Rise = 40, run = 1, slope = $\frac{40}{1} = 40$

   c. Line descends from left to right, negative slope. Rise = -20, run = 1, slope = $\frac{-20}{1} = -20$

7. Both x and y increase, positive slope.

   Slope = $\frac{3-2}{4-0} = \frac{1}{4}$

9. Line goes from Quadrant 2 to Quadrant 4, negative slope.

   Slope = $\frac{-4-3}{0-(-2)} = \frac{-7}{2} = -3.5$

11. x value does not change, vertical line, undefined slope.

    Slope = $\frac{4-3}{4-4} = \frac{1}{0}$, *undefined*

13. y value does not change, horizontal line, zero slope.

    Slope = $\frac{2-2}{-2-0} = \frac{0}{-2} = 0$

15. Line goes from Quadrant 2 to Quadrant 4, negative slope.

    Slope = $\frac{-1-3}{4-(-2)} = \frac{-4}{6} = \frac{-2}{3}$

17. y value does not change, horizontal line, zero slope.

    Slope = $\frac{-3-(-3)}{4-2} = \frac{0}{2} = 0$

19. Line goes from positive vertical axis to positive horizontal axis, negative slope.

    Slope = $\frac{0-4}{5-0} = \frac{-4}{5}$

Section 4.3 (con't)

21. x value does not change, vertical line, undefined slope.

$$\text{Slope} = \frac{4-(-4)}{3-3} = \frac{8}{0}, \text{ undefined}$$

23. As x values increase y values decrease, negative slope. Δx between each input is 1, Δy between each output is -3, linear function. Slope = -3, no units are given.

25. As x values increase y values increase, positive slope. Δx between each input is 1, Δy between each output is 7, linear function. Slope = 7, no units are given.

27. As x values increase y values increase, positive slope. Δx between each input is 2, Δy between each output is 18, linear function. Slope = $\frac{18}{2}$ = 9. Units are

$$\frac{\text{earnings}}{\text{hour}}.$$

29. As x values increase y values increase, positive slope. Δx between each input is 1, Δy between each output is $0.50, linear function. Slope = $0.50. Units are $\dfrac{\text{cost}}{\text{kilogram}}$.

31. As x values increase y values increase, positive slope. Δx between each input is 2, Δy between each output is $0.64, linear function. Slope $= \dfrac{\$0.64}{2} = \$0.32$.

Units are $\dfrac{\text{cost}}{\text{pound}}$.

33. As x values increase y values increase, positive slope. Δx between each input is 1, Δy between each output varies, non-linear function. Units are $\dfrac{\text{ft}}{\text{sec}}$.

35. As x values increase y values decrease, negative slope. Δx between each input and Δy between each output appear to vary, a comparison of slope between each point reveals a linear function.

$$\text{Slope} = \frac{-\$2.50}{10} = \frac{-\$1.25}{5} = -\$0.25.$$

Units are $\dfrac{\$ \text{ value}}{\text{copy}}$.

37. As x values increase y values increase, positive slope. Δx between each input is 1, Δy between each output varies, non-linear function. Units are $\dfrac{\text{miles}}{\text{hour}}$.

39. $\dfrac{y_2 - y_1}{x_2 - x_1} = \dfrac{d-b}{c-a}$

Section 4.3 (con't)

41. $\dfrac{y_2 - y_1}{x_2 - x_1} = \dfrac{b - 0}{0 - a} = -\dfrac{b}{a}$

43. $\dfrac{y_2 - y_1}{x_2 - x_1} = \dfrac{c - b}{a - a} = \dfrac{c - b}{0}$, *undefined*

45.

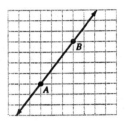

47.

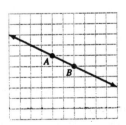

49.

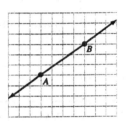

51.

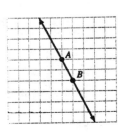

53.

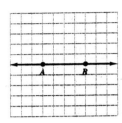

55.

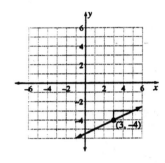

57.

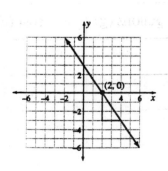

59.

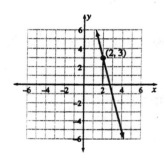

61.

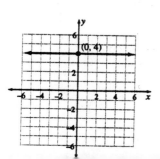

Section 4.3 (con't)

63.

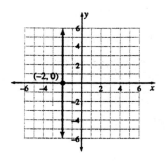

65. $\frac{3}{4}, 1, \frac{4}{3}$

67. $0, -\frac{1}{2}, -\frac{2}{1}$

69. $\frac{3}{5}, \frac{3}{4}, 1, \frac{3}{2}$

71. Rise = 5 ft, run = 10 ft

Slope = $\dfrac{5\,\text{ft}}{10\,\text{ft}} = \dfrac{1}{2}$

73. Rise = 7 ft, run = $\dfrac{28\,\text{ft}}{2} = 14\,\text{ft}$

Slope = $\dfrac{7\,\text{ft}}{14\,\text{ft}} = \dfrac{1}{2}$

75. a.

gallons (g)	cost (c)
0	0
1	$1.55
2	2($1.55) = $3.10

b. Slope = $1.55 $\dfrac{\text{cost}}{\text{gallon}}$

c. c = $1.55g

77. a.

hours (h)	earnings (e)
0	0
1	$6.25
2	2($6.25) = $12.50

b. Slope = $6.25 $\dfrac{\$\,\text{earnings}}{\text{hour}}$

c. e = $6.25h

79. a.

hours (x)	distance (d) in kilometers
0	0
1	80
2	2(80) = 160

b. Slope = 80 $\dfrac{\text{km}}{\text{hr}}$

c. d = 3x

81. a.

hours (x)	cost (c)
0	$3
1	$1(1) + $3 = $4
2	$1(2) + $3 = $5

b. Slope = cost per hour = $1

c. c = $1x + $3

Section 4.3 (con't)

83. a.

meal (x)	total cost (c)
0	0
$1	$1 + 0.15($1) = $1.15
$2	$2 + 0.15($2) = $2.30

  b. Slope = 1.15 $\dfrac{\text{total cost}}{\text{meal cost}}$

  c. c = 1.15x

85. Because of the constant fee in each exercise.

87. Freezing ordered pair (0, 32)

   Boiling ordered pair (100, 212)

   Slope = $\dfrac{212 - 32}{100 - 0} = \dfrac{180}{100} = \dfrac{9}{5}$ $\dfrac{°F}{°C}$

89. Slope tells you relative direction up or down from left to right and how steep the graph is as it climbs or descends.

91. Slope will be negative when the y values get smaller as the x values get larger.

93. Going left to right on the graph, count the change in y and divide by the change in x. Write the appropriate sign.

95. From a point, go *a* units up if *a* is positive and down if *a* is negative and go *b* units horizontally to the right to find a second point. Draw a line between the two points.

97. The answer shows the change in x over the change in y. Slope is the reciprocal; change in y over the change in x.

99. a.

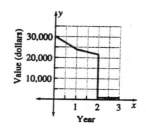

  b. $\dfrac{30,000 - 24,000}{0 - 1}$ = -$6,000

  c. $\dfrac{24,000 - 21,600}{1 - 2}$ = -$2,400

  d. Extreme loss of value happened suddenly.

  e. $\dfrac{1000 - 1000}{2 - 3}$ = 0

  f. Answers will vary, after driving it for 2 years, she had a wreck in the third year.

Section 4.4

1. Slope = -1, y-intercept = 1

   y = -x + 1

3. Slope = 2, y-intercept = 2

   y = 2x + 2

5. Slope = 0, y-intercept = -1

   y = -1

7. Slope = $\frac{1}{30}$, C-intercept = 1

   C = $\frac{1}{30}$ n + 1

9. Slope = $\dfrac{-200}{50} = -4$,

   C-intercept = 160

   C = -4t + 160

11. $m = 2$, $b = -\frac{1}{2}$

13. $m = -4$, $b = 15$

15. $m = -\frac{3}{4}$, $b = 0$

17. $2x = y + 4$,

    $y = 2x - 4$,

    $m = 2$, $b = -4$

19. $2x + 3y = 12$,

    $3y = -2x + 12$,

    $y = -\frac{2}{3}x + 4$

    $m = -\frac{2}{3}$, $b = 4$

21. $5y - 2x = 10$,

    $y = 2x + 10$,

    $y = \frac{2}{5}x + 2$

    $m = \frac{2}{5}$, $b = 2$

23. $x - 4y = 4$,

    $4y = x - 4$,

    $y = \frac{1}{4}x - 1$

    $m = \frac{1}{4}$, $b = -1$

25. $m = -0.30$, $b = 12$

27. $m = 55$, $b = 0$

29. $m = 2\pi$, $b = 8$

31. $m = 2.98$, $b = 0.50$

33. $m = -0.29$, $b = 50$

35. H = 0.8(200 - A) = 160 - 0.8A

    $m = -0.8$, $b = 160$

37. C = 65 + 0.15(d - 100)

    C = 65 + 0.15d - 15 = 0.15d + 50

    $m = 0.15$, $b = 50$

39. $y = \frac{1}{2}x + 3$

41. $y = \frac{2}{3}x - 2$

43. $y = 5x + \frac{1}{4}$

45. $y = -\frac{3}{2}x + 1$

Section 4.4 (con't)

47. $-1 = 4(3) + b,$

$-1 - 12 = b,$

$b = -13$

$y = 4x - 13$

49. $-2 = (-1)(4) + b,$

$-2 + 4 = b,$

$b = 2$

$y = -x + 2$

51. $4 = (\frac{1}{2})(2) + b,$

$4 - 1 = b,$

$b = 3$

$y = \frac{1}{2}x + 3$

53. $3 = (\frac{4}{5})(-10) + b,$

$3 + 8 = b,$

$b = 11$

$y = \frac{4}{5}x + 11$

55. $1 = (\frac{5}{3})(-3) + b,$

$1 + 5 = b,$

$b = 6$

$y = \frac{5}{3}x + 6$

57. $3 = (-2)(1.5) + b,$

$3 + 3 = b,$

$b = 6$

$y = -2x + 6$

59. $m = \dfrac{9-1}{3-1} = \dfrac{8}{2} = 4$

$1 = (4)(1) + b,$

$1 - 4 = b,$

$b = -3$

$y = 4x - 3$

61. $m = \dfrac{-8-(-2)}{5-2} = \dfrac{-6}{3} = -2$

$-2 = (-2)(2) + b,$

$-2 + 4 = b,$

$b = 2$

$y = -2x + 2$

63. $m = \dfrac{-1-1}{0-(-3)} = \dfrac{-2}{3}$

$b = -1, \text{ from } (0, -1)$

$y = -\frac{2}{3}x - 1$

Section 4.4 (con't)

65. $m = \dfrac{0-6}{10-13} = \dfrac{-6}{-3} = 2$

$6 = 2(13) + b,$

$6 - 26 = b,$

$b = -20$

$y = 2x - 20$

67. $m = \dfrac{-2-6}{-4-(-5)} = \dfrac{-8}{1} = -8$

$6 = (-8)(-5) + b,$

$6 - 40 = b,$

$b = -34$

$y = -8x - 34$

69. $m = \dfrac{3-2}{3-5} = \dfrac{1}{-2}$

$2 = -\frac{1}{2}(5) + b,$

$2 + \frac{5}{2} = b,$

$b = 4\frac{1}{2}$

$y = -\frac{1}{2}x + 4\frac{1}{2}$

71. d. $2y = x + 4,\ y = \frac{1}{2}x + 2$

a and c have slopes that multiply to -1 so they are perpendicular.

b and d have the same slope so they are parallel.

73. b. $x - \frac{1}{3}y = 6,$

$x - 6 = \frac{1}{3}y,$

$y = 3x - 18$

c. $3y - x = 2,$

$3y = x + 2,$

$y = \frac{1}{3}x + \frac{2}{3}$

d. $y + \frac{1}{3}x = 4,$

$y = -\frac{1}{3}x + 4$

a and c have the same slope so they are parallel.

b and d have slopes that multiply to -1 so they are perpendicular.

75.

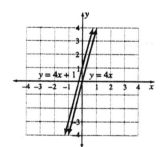

$b = 0,\ y = 4x$

77.

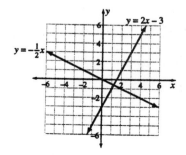

$b = 0,\ 2 \cdot m = -1,\ m = -\frac{1}{2}$

$y = -\frac{1}{2}x$

Section 4.4 (con't)

79.

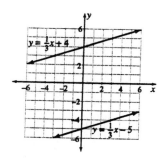

$b = 4,\ m = \frac{1}{3}$

$y = \frac{1}{3}x + 4$

81.

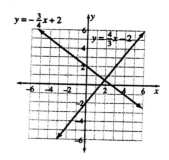

$b = -2,\ \left(-\frac{3}{4}\right)m = -1,\ m = \frac{4}{3}$

$y = \frac{4}{3}x - 2$

83.

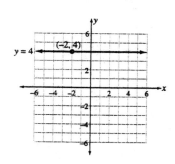

$m = 0 \, (horizontal\ line)$

$y = 4$

85.

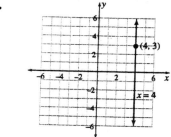

m is undefined $(vertical\ line)$,

no y - intercept

$x = 4$

87.

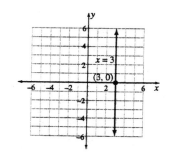

m is undefined $(vertical\ line)$,

no y - intercept

$x = 3$

89.

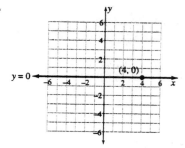

$m = 0,\ b = 0,\ y = 0$ (x-axis)

Section 4.4 (con't)

91. a. Rental per hour, -$2

  b. Prepaid amount, $50

  c. $y = -\$2x + \$50$

93. a. Rental per hour, $42

  b. Insurance fee, $28

  c. $y = \$42x + \$28$

95. per hour, percent of

97. Slope did not change; parallel line.

  C = $0.25n + $45

99. Slope increased; steeper line.

  C = $1.60g

101. Slope did not change; parallel line.

  C = 0.01x + $5

103. Data points are (7, $6.01), (8, $6.79)

  $$m = \frac{y_2 - y_1}{x_2 - x_1} = \frac{6.79 - 6.01}{8 - 7} = \$0.78$$

  $b = y - mx,$

  $b = \$6.01 - \$0.78(7),$

  $b = \$0.55$

  $y = \$0.78x + \$0.55$

105. a. AB $\quad m = \dfrac{20 - 0}{20 - 0} = 1$

  BC $\quad m = \dfrac{20 - 20}{200 - 20} = 0$

CD $\quad m = \dfrac{50 - 20}{500 - 200} = \dfrac{30}{300} = \dfrac{1}{10}$

DE $\quad m = \dfrac{150 - 50}{600 - 500} = \dfrac{100}{100} = 1$

  b. AB $\quad b = 0$; yes

  BC (horizontal line) $\quad b = 20$; no

  CD $\quad b = 50 - \frac{1}{10}(500), b = 0$; no

105. **b.** DE $\quad b = 150 - (1)(600),$

  $b = -450$; no

  c. AB $\quad y = x$

  BC $\quad y = 20$

  CD $\quad y = \frac{1}{10}x$

  DE $\quad y = x - 450$

  d. AB $\quad 0 \le x \le 20$

  BC $\quad 20 < x \le 200$

  CD $\quad 200 < x \le 500$

  DE $\quad 500 < x \le 600$

  e. AB $\quad 0 \le y \le 20$

  BC $\quad y = 20$

  CD $\quad 20 < y \le 50$

  DE $\quad 50 < y \le 150$

  AB and DE are parallel, they have the same slope.

Section 4.4 (con't)

107. Find the slope from $m = \dfrac{y_2 - y_1}{x_2 - x_1}$, then

substitute one ordered pair and *m* into

$b = y - mx$ to find *b*. Place *m* and *b* into

$y = mx + b$.

109. c = slope, d = y-intercept.

111. Slope $= -\frac{b}{a}$

113. a. $m = \dfrac{b - 0}{0 - a} = -\dfrac{b}{a}$

    b. The opposite of the y-intercept
divided by the x-intercept.

    c. $m = -\dfrac{-4}{-5} = -\dfrac{4}{5}$

    $m = \dfrac{-4 - 0}{0 - (-5)} = \dfrac{-4}{5} = -\dfrac{4}{5}$

Section 4.5

1. Multiply both sides by -1 and reverse the inequality sign.

3. Multiply both sides by -2 and reverse the inequality sign.

5. Subtract 2x from both sides, divide both sides by -3 and reverse the inequality sign.

7. $-2 + 3 = 1$, $1 < 3$, $(-2, 3)$ is not a solution

   $4 + 0 = 4$, $4 > 3$, $(4, 0)$ is a solution

   $1 + 4 = 5$, $5 > 3$, $(1, 4)$ is a solution

9. $\frac{1}{2}(-2) + 3 = 2$, $2 \geq 2$, $(-2, 3)$ is a solution.

   $\frac{1}{2}(4) + (-2) = 0$, $0 < 2$, $(4, -2)$ is not a solution

   $\frac{1}{2}(6) + (-1) = 2$, $2 \geq 2$, $(6, -1)$ is a solution

11. $2(0) - 3(0) = 0$, $0 < 4$, $(0, 0)$ is not a solution

    $2(1) - 3(-1) = 5$, $5 > 4$, $(1, -1)$ is a solution

    $2(3) - 3(0) = 6$, $6 > 4$, $(3, 0)$ is a solution

13. $3 > -2$, $(-2, 3)$ is not a solution

    $0 > -2$, $(-2, 0)$ is not a solution

    $-2 \leq -2$, $(0, -2)$ is a solution

15. For x-intercept;

    $0 = -3x + 2$, $3x = 2$, $x = \frac{2}{3}$

    for y-intercept; $y = 2$

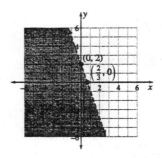

17. For x-intercept;

    $0 = 4x - 1$, $4x = 1$, $x = \frac{1}{4}$

    For y-intercept; $y = -1$

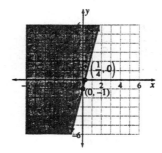

19. For x-intercept;

    $0 = 2 - 2x$, $2x = 2$, $x = 1$

    For y-intercept; $y = 2$

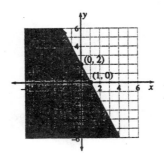

Section 4.5 (con't)

21. For x-intercept;

$2x + 0 = 5$, $2x = 5$, $x = \frac{5}{2}$

For y-intercept; $y = 5$

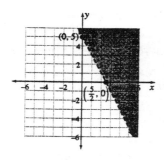

23. For x-intercept;

$2 - 2(0) = 4x$, $4x = 2$, $x = \frac{1}{2}$

For y-intercept;

$2 - 2y = 4(0)$, $2 = 2y$, $y = 1$

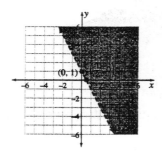

25. x-intercept = 4, no y-intercept

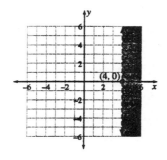

27. No x-intercept, y-intercept = 4

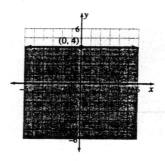

29. Boundary line is $y = 3$

Inequality is $y < 3$

31. Boundary line is $x = 3$

Inequality is $x \leq 3$

33. Boundary line is $y = x + 4$

Inequality is $y \leq x + 4$

35. Boundary line is $y = 2x + 1$

Inequality is $y > 2x + 1$

37. Test an ordered pair in the inequality, if true shade the side containing the point, if false shade the other side.

39. The first is a line graph with a dot at x = 3 and an arrow to the right. The second has a vertical boundary line x = 3 and shading to the right.

Section 4.5 (con't)

41. $16x + 12y \geq 2400$

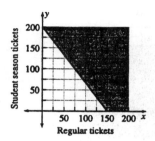

43. a. Could include 0 apricots and 4
tangerines (0, 4); 7 apricots and 0
tangerines (7, 0); 3 apricots and 2
tangerines (3, 2).

b.

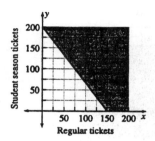

c. Solution sets are positive numbers

$20a + 35t \leq 140, a \leq 7, t \leq 4$

45. a. 4 quarters = \$1, \$2 - \$1 = \$1

\$1 = 10 dimes, ordered pair is (10, 4)

With 0 quarters the entire \$2 is in dimes

\$2 = 20 dimes, ordered pair is (20, 0)

45. b. 15 dimes = \$1.50,

\$2 - \$1.50 = \$0.50 = 2 quarters, ordered
pair is (15, 2)

With 0 dimes the entire \$2 is in quarters

\$2 = 8 quarters, ordered pair is (0, 8)

c. \$0.10x

d. \$0.25y

e. \$0.10x + \$0.25y \leq \$2

f. Domain: $0 \leq x \leq 20$

Range: $0 \leq y \leq 8$,

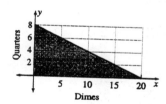

47. 4th Quadrant including positive x-axis.

49. 3rd Quadrant

51. x > 0, y < 0

Chapter 4 Review

1. $6 = 4(-1) + b$

$6 = -4 + b, \quad b = 10$

3. $37 = \frac{5}{9}(F - 32)$

$\frac{9}{5}(37) = F - 32, \quad 66.6 + 32 = F$

$F = 98.6$

5. Area in terms of base and height.

7. $W = hp,$

$\frac{W}{p} = \frac{hp}{p},$

$h = \frac{W}{p}$

9. $A = \frac{bh}{2},$

$2A = 2 \cdot \frac{bh}{2},$

$\frac{2A}{h} = \frac{bh}{h},$

$b = \frac{2A}{h}$

11. $I = \frac{AH}{T},$

$I \cdot T = T \cdot \frac{AH}{T},$

$\frac{IT}{I} = \frac{AH}{I},$

$T = \frac{AH}{I}$

13. $PV = nRT,$

$\frac{PV}{nR} = \frac{nRT}{nR},$

$T = \frac{PV}{nR}$

15. $ax + by = c,$

$ax = c - by$

$x = \frac{c - by}{a}$

17. $C = 35 + 5(k - 100),$

$C - 35 = 5(k - 100)$

$\frac{C}{5} - 7 = k - 100,$

$k = \frac{C}{5} + 93$

19. $P_1 V_1 = P_2 V_2,$

$\frac{P_1 V_1}{V_2} = \frac{P_2 V_2}{V_2},$

$P_2 = \frac{P_1 V_1}{V_2}$

21. $A = \frac{1}{2} h(b_1 + b_2),$

$2A = h(b_1 + b_2),$

$\frac{2A}{h} = b_1 + b_2,$

$b_1 = \frac{2A}{h} - b_2$

Chapter 4 Review (con't)

23. $27 = \frac{1}{2}(3)(9 + b)$

Using solution to exercise #21:

$b = \dfrac{2(27)}{3} - 9 = 18 - 9 = 9$

25. $f(x) = 7 - 2x$

$f(0) = 7 - 2(0) = 7$

$f(3) = 7 - 2(3) = 7 - 6 = 1$

$f(-5) = 7 - 2(-5) = 7 + 10 = 17$

$f(a) = 7 - 2a$

27. $f(x) = x^2 + x$

$f(0) = 0^2 + 0 = 0$

$f(3) = 3^2 + 3 = 9 + 3 = 12$

$f(-5) = (-5)^2 + (-5) = 25 - 5 = 20$

$f(a) = a^2 + a$

29. $f(x) = 0$,

$2x - 5 = 0$,

$2x = 5$,

$x = \frac{5}{2} = 2\frac{1}{2}$

31. $f(x) = 0$

$-3x - 4 = 0$,

$-3x = 4$,

$x = -\frac{4}{3} = -1\frac{1}{3}$

33. $f(x) = 0$,

$\frac{1}{2}x - 6 = 0$,

$\frac{1}{2}x = 6$,

$x = 12$

35. $f(0)$ is the y-intercept.

37. a. Inputs are x values, {2, 4, 6, 8}

b. Outputs are y values

{25, 50, 75, 100}

c. The value of "bits" in cents.

39. a. Domain is x values,

{1, 2, 3, 4, 5, 6}

b. Range is y values,

{2, 6, 10, 20, 40, 60}

c. The name of the nail in "pennies", usually abbreviated d.

41. One output for each input, describes a function.

43. One output for each input, describes a function.

45. Many outputs for same input, not a function.

47. b. Substitute y = 0 into the equation and solve for x.

49. a. Let x = a.

Chapter 4 Review (con't)

51. a. $\dfrac{4-(-2)}{6-4} = \dfrac{6}{2} = 3$

b. $\dfrac{3-(-4)}{-6-4} = \dfrac{7}{-10} = -\dfrac{7}{10}$

c. $\dfrac{0-(-2)}{6-(-4)} = \dfrac{2}{10} = \dfrac{1}{5}$

d. $\dfrac{1-3}{-6-9} = \dfrac{-2}{-15} = \dfrac{2}{15}$

53.

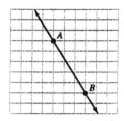

55.

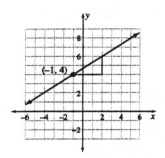

57. Line goes down from left to right.

59. Locate (x, y), count a units up (or down) and b units to the right and make a second point. Draw the line between the points.

61. Rise = 3, run = 1; $m = \frac{3}{1} = 3$, $b = 2$

$y = 3x + 2$

63. Rise = -1, run = 2; $m = -\frac{1}{2}$, $b = 1$

$y = -\frac{1}{2}x + 1$

65. Δx is 2 between each input, Δy is -3 between each output, linear function. Slope $m = \frac{-3}{2}$.

67. Δx is 2 between each input, Δy varies, non-linear function.

69. a. Let x = minutes and y = cost

(14, 4.44), (5, 2.10)

b. $m = \dfrac{4.44 - 2.10}{14 - 5} = \dfrac{2.34}{9} = 0.26$

c. Slope is cost per minute.

71. a. Let x = hours and y = feet

(12, 12), (17, 32)

b. $m = \dfrac{32 - 12}{17 - 12} = \dfrac{20}{5} = 4$

c. Slope is feet per hour.

73. Solving y + b = mx for y gives

y = mx - b, not equivalent to the other equations.

75. 3x + 5y = 15,

5y = -3x + 15

$y = -\frac{3}{5}x + 3$,

$m = -\frac{3}{5}$, $b = 3$

Chapter 4 Review (con't)

77. 5x - 2y = 10,

$-2y = -5x + 10$

$y = \frac{5}{2}x - 5,$

$m = \frac{5}{2}, b = -5$

79. 4y - 3x = 8,

$4y = 3x + 8$

$y = \frac{3}{4}x + 2,$

$m = \frac{3}{4}, b = 2$

81. y + 3 = 0

$y = -3,$

$m = 0, b = -3$

83. $x = 3,$

Slope is undefined, there is no y-intercept.

85. c = 2 + 1.5(n - 1)

$c = 2 + 1.5n - 1.5$

$c = 1.5n + 0.5$

$m = 1.5, \ b = 0.5$

87. $y = 2x + 4$

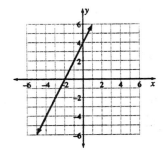

89. $y = 3x$

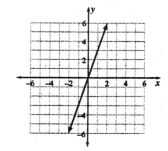

91. $y = \frac{1}{2}x + 2, \ 2y = x + 4$

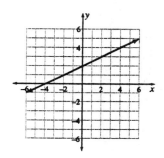

93. $y = -3x + \frac{1}{4}, \ 4y = -12x + 1$

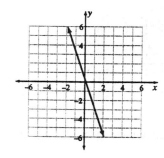

Chapter 4 Review (con't)

95. x = 4

97. x = -2

99. y = -2

101. y = 0 (x-axis)

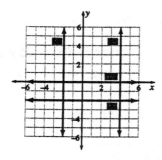

103. Parallel line has same slope, b = 0

$y = 3x$

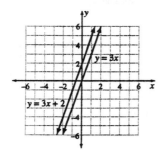

105. Perpendicular line has slope $m = \frac{3}{2}$,

b = -2. $y = \frac{3}{2}x - 2$

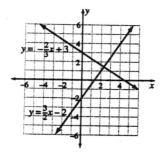

107. To find if points lie on a straight line, check the slope between pairs of points.

$$m_1 = \frac{3-4}{0-(-2)} = \frac{-1}{2},$$

$$m_2 = \frac{1-4}{4-(-2)} = \frac{-3}{6} = \frac{-1}{2}$$

Points lie on a straight line, b = 3 is given by (0, 3).

$y = -\frac{1}{2}x + 3$

109. Activity factor = 10, under age 35

C = 10w

111. Δx is 1 and Δy is 175 between pairs of data, therefore data are linear. Use any data point to solve for b.

507 = 175(1) + b, b = 332

y = 175x + 332, where y = income and

x = number of people

113. a. 2 ≤ 4 - 3, false

b. -6 ≤ -1 - 3, true

c. 8 ≤ 8 - 3, false

d. 0 ≤ 0 - 3, false

115. a. y ≥ 0

b. x ≥ 0

c. x < 0

d. y < 0

Chapter 4 Review (con't)

115. e. $x > 0$

   f. $y \leq 0$

117.

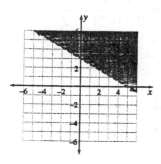

119.

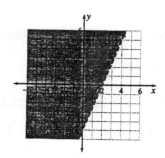

121.

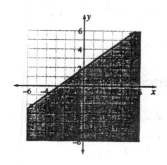

123. Let x = green olives and y = black olives. Combinations will vary, those that have fewer than 400 calories will satisfy $20x + 25y < 400$, those with exactly 400 calories will satisfy

$20x + 25y = 400$.

Combined equation is $20x + 25y \leq 400$

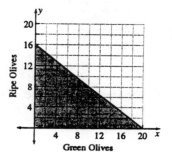

Chapter 4 Test

1. $b = 5 - \frac{1}{2}(-8)$,

   $b = 9$

2. Answers will vary but could include:

   $$G = \frac{T_1 + T_2 + H + F}{\text{Total Points}}$$

3. **a.** $C = \pi d$,

   $$\frac{C}{\pi} = \frac{\pi d}{\pi},$$

   $$d = \frac{C}{\pi}$$

   b. $A = \frac{1}{2}bh$,

   $2A = bh$,

   $$h = \frac{2A}{b}$$

   c. $y = mx + b$,

   $b = y - mx$

   d. $P_1V_1 = P_2V_2$,

   $$\frac{P_1V_1}{P_2} = \frac{P_2V_2}{P_2},$$

   $$V_2 = \frac{P_1V_1}{P_2}$$

4. **a.** One output for each input, function.

   b. Many outputs for one input, not a function.

   c. One output for each input, function.

   d. One output for each input, function.

   e. Many outputs for one input, not a function.

   f. One output for each input, function.

5. **a.** $f(x) = 0$, ($y = 0$ is the x-axis)

   b. $f(0)$, ($x = 0$ is the y-axis)

   c. $f(a)$

6. **a.** (1) Horizontal line, $m = 0$

   (2) Rise = 50, run = 5, $m = \frac{50}{5} = 10$

   (3) Rise = 300, run = 6, $m = \frac{300}{6} = 50$

   (4) Vertical line, slope is undefined

   b. (2), slope = 10

7. **a.** $m = \dfrac{-1-(-3)}{2-5} = \dfrac{2}{-3} = -\dfrac{2}{3}$

   b. $m = \dfrac{-4-(-4)}{0-1} = \dfrac{0}{-1} = 0$

   c. $m = \dfrac{-4-3}{2-1} = \dfrac{-7}{1} = -7$

   d. $m = \dfrac{1-(-3)}{4-4} = \dfrac{4}{0}, \textit{undefined}$

Chapter 4 Test (con't)

8.

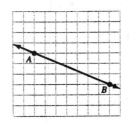

9.

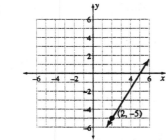

10. a. $y = 5 - 2x$, $m = -2$, $b = 5$

b. $6y + 3 = 2x$,

$6y = 2x - 3$

$y = \frac{2}{6}x - \frac{3}{6}$,

$y = \frac{1}{3}x - \frac{1}{2}$, $m = \frac{1}{3}$, $b = -\frac{1}{2}$

11. a. Different slopes, same y-intercept, neither.

b. Same slope, different y-intercept, parallel.

c. Slopes are opposite reciprocals, perpendicular.

d. Reciprocal slopes, different y-intercepts, neither

12. Substitute slope and ordered pair into

$b = y - mx$ and solve for b. Put b and m into $y = mx + b$.

13. Parallel line has same slope, *a* and given y-intercept, *k*. $y = ax + k$.

14. For a horizontal line $y = k$.

15. $y = 5x - 1$

16. $y = \frac{1}{3}x + 2$, $3y = x + 6$

17.

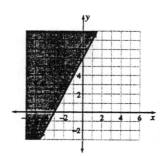

18.

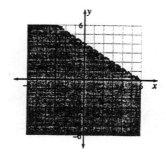

19. a. x, Miles

b. y, Cost

c. From the table, 125 is between 100 & 150, cost will be between 40 & 45, from the graph cost is $42.50.

Chapter 4 Test (con't)

19. d. From the table it appears that every
10 miles costs an additional $1.
Extending the table $57 will buy 270
miles. This is confirmed on the graph.

  e. $30 from the graph. Base price to
rent the vehicle.

  f. $\Delta x = 50$, $\Delta y = 5$, slope $= \frac{5}{50} = 0.10$,
the cost per mile.

  g. $y = 0.10x + 30$

  h. $0 = 0.10x + 30$, $-30 = 0.10x$

$x = -300$, has no meaning

Cumulative Review Chapters 1 to 4

1. $\frac{1}{3} + \frac{3}{7} = \frac{7}{21} + \frac{9}{21} = \frac{16}{21}$

$\frac{1}{3} - \frac{3}{7} = \frac{7}{21} - \frac{9}{21} = -\frac{2}{21}$

$\frac{1}{3} \cdot \frac{3}{7} = \frac{3}{21} = \frac{1}{7}$

$\frac{1}{3} \div \frac{3}{7} = \frac{1}{3} \cdot \frac{7}{3} = \frac{7}{9}$

3. $4.8 + (-6.4) = 4.8 - 6.4 = -1.6$

$4.8 - (-6.4) = 4.8 + 6.4 = 11.2$

$4.8(-6.4) = -30.72$

$4.8 \div (-6.4) = -0.75$

5. $4x^2 + 6x$

$4x^2 - 6x$

$4x^2(6x) = (4)(6)(x^2)(x) = 24x^3$

$4x^2 \div 6x = \dfrac{4x^2}{6x} = \frac{2}{3}x$

7. $1{,}000{,}000 \text{ min} \cdot \dfrac{1\,\text{hr}}{60\,\text{min}} \cdot \dfrac{1\,\text{day}}{24\,\text{hr}} \approx$

694.4 days

9. $A = 91 \text{ in}^2$, $(a + b) = 13$ in

$91 = \frac{1}{2}h(13)$,

$182 = 13h$,

$h = 14$ in

11. $2(x + 5) = x + 3$

$2x + 10 = x + 3$, $x + 10 = 3$

$x = -7$

13. $9 - 2(x - 3) = 21$, $-2(x - 3) = 12$

$x - 3 = -6$, $x = -3$

15. a. $m = -\frac{3}{2}$

b. $b = 9$

17.

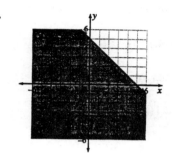

19. $f(-2) = (-2)^2 = 4$

21. $m = \dfrac{-6 - 0}{2 - (-4)} = \dfrac{-6}{6} = -1$

$b = y - mx$, $b = 0 - (-1)(-4) = -4$

$y = -x - 4$

Section 5.1

1. a. $\frac{5}{3}$, $5:3$

b. $3:2$, 3 to 2

c. $\frac{4}{9}$, 4 to 9

3. a. Common factor is 5

$$\frac{15}{35} = \frac{5\cdot 3}{5\cdot 7} = \frac{3}{7}$$

b. Common factor is 16

$$\frac{48}{16} = \frac{16\cdot 3}{16\cdot 1} = \frac{3}{1}$$

c. Common factor is 240

$$\frac{5280}{3600} = \frac{240\cdot 22}{240\cdot 15} = \frac{22}{15}$$

5. a. Common factors are 4x,

$12x : 4x^2 = 3 : x$

b. Common factors are 2xy,

2xy to $6x^2y$ = 1 to 3x

c. Common factor is x,

$x : 3x^4 = 1 : 3x^3$

7. a. Common factors are ac,

abc to ace = b to e

b. Common factor is a,

a(b + c) to ac = (b + c) to c

c. Common factor is (a + b),

(a + b) : a(a + b) = 1 : a

9. Slope = 3, slope ratio is $\frac{3}{1}$.

11. Slope = -4, slope ratio is $\frac{-4}{1}$.

13. Slope and slope ratio = $\frac{1}{2}$.

15. Slope = 2.98, slope ratio = $\frac{2.98}{1}$ or $\frac{149}{50}$

17. Slope = -0.05, slope ratio = $\frac{-0.05}{1}$ or $-\frac{1}{20}$

19. a. $\frac{3}{15} = \frac{3\cdot 1}{3\cdot 5} = \frac{1}{5}$

b. $\frac{7.5}{10} = \frac{2.5\cdot 3}{2.5\cdot 4} = \frac{3}{4}$

c. $\frac{6.75}{10.5} = \frac{0.75\cdot 9}{0.75\cdot 14} = \frac{9}{14}$

d. $\frac{7.25}{9.75} = \frac{0.25\cdot 29}{0.25\cdot 39} = \frac{29}{39}$

21. a. $209 \div 585 = 0.357$

b. $152 \div 584 = 0.260$

23. Let x = hourly rate;

$4x + 3x + 1x = \$400$

$8x = \$400$, x = \$50

A = 4x = 4($50) = \$200

B = 3x = 3($50) = \$150

C = 1x = \$50

Section 5.1 (con't)

25. Continued ratio of 4 : 3 : 3 is a total of

10 shares, $2500 \div 10 = 250$ per share,

Cosmo gets $4(250) = 1000$ crowns

Timo and Sven get $3(250) = 750$ crowns

each.

27. 9 carbon, 8 hydrogen, 4 oxygen

29. 1 calcium, 1 carbon, 3 oxygen

31. $62.5\% = \dfrac{62.5}{100} = \dfrac{5 \cdot 12.5}{8 \cdot 12.5} = \dfrac{5}{8}$

33. $\dfrac{1}{12} = 0.083\overline{3} = 8\frac{1}{3}\%$

35. $\dfrac{1}{8} = 12.5\%$

37. Total income is $1200 (sum of all

expenditures)

$\dfrac{360}{1200} = 30\%,\ \dfrac{132}{1200} = 11\%,\ \dfrac{240}{1200} = 20\%,$

$\dfrac{84}{1200} = 7\%,\ \dfrac{96}{1200} = 8\%,\ \dfrac{48}{1200} = 4\%,$

$\dfrac{60}{1200} = 5\%,\ \dfrac{24}{1200} = 2\%,\ \dfrac{156}{1200} = 13\%$

39. $\dfrac{16 - 12}{12} = \dfrac{4}{12} = \dfrac{1}{3} = 33.3\%$ change

$\dfrac{16}{12} = \dfrac{4}{3} = 1\frac{1}{3} = 133.3\%$

41. $\dfrac{0.79 - 1.00}{1.00} = \dfrac{-0.21}{1} = -21\%$ change

$\dfrac{0.79}{1.00} = 0.79 = 79\%$

43. a. Let x be the new rent;

$\dfrac{x - 450}{450} = 0.20,\ x - 450 = 90$

$x = \$540$

b. Let x be the old rent;

$\dfrac{450 - x}{x} = 0.20,\ 450 - x = 0.20x$

$450 = 1.20x,\ x = \$375$

45. a. $\frac{1}{2}$ foot $= 6$ inches, ratio is 6 to 1

b. 3000 grams $\cdot \dfrac{1 \text{ kilogram}}{1000 \text{ grams}} = 3 \text{ kg}$

ratio is 3 to 1

c. 32 ounces $\cdot \dfrac{1 \text{ pound}}{16 \text{ ounces}} = 2 \text{ pounds}$

ratio is 2 to 5

d. $2\frac{2}{3}$ yards $\cdot \dfrac{3 \text{ feet}}{1 \text{ yard}} = 8 \text{ feet}$

ratio 8 to 2 = 4 to 1

e. 20 liters $\cdot \dfrac{1000 \text{ ml}}{1 \text{ liter}} = 20{,}000 \text{ ml}$

ratio 200 to 20,000 = 1 to 100

f. 2 years $= 24$ months

ratio 24 to 150 = 4 to 25

g. $\frac{1}{4}$ hour $= 15$ minutes

ratio 20 to 15 = 4 to 3

Section 5.1 (con't)

47. 1 mile = 5280 feet

30 feet per mile $= \dfrac{30}{5280} \approx 0.006 \approx 0.6\%$

49. Answers will vary

51. 3 to 1

53. \$480 to 12 = \$40 to 1

55. 1 gallon = 8 pints,

   0.5 to 8 = 1 to 16

57. $\dfrac{240 \text{ cc}}{16 \text{ hours}} = 15$ cc per hour

59. $\dfrac{30 \text{ miles}}{\text{hour}} \cdot \dfrac{5280 \text{ feet}}{1 \text{ mile}} \cdot \dfrac{1 \text{ hour}}{60 \text{ min}} \cdot \dfrac{1 \text{ min}}{60 \text{ sec}}$

   = 44 feet per second

61. $\dfrac{80 \text{ miles}}{\text{hour}} \cdot \dfrac{5280 \text{ feet}}{1 \text{ mile}} \cdot \dfrac{1 \text{ hour}}{60 \text{ min}} \cdot \dfrac{1 \text{ min}}{60 \text{ sec}}$

   $= 117\frac{1}{3}$ feet per second

63. $\dfrac{88 \text{ ft}}{\text{sec}} \cdot \dfrac{60 \text{ sec}}{1 \text{ min}} \cdot \dfrac{60 \text{ min}}{1 \text{ hour}} \cdot \dfrac{1 \text{ mile}}{5280 \text{ ft}}$

   = 60 miles per hour

   BACKWARD

   $\dfrac{88}{5280} \times 3600$

65. $\dfrac{220 \text{ ft}}{\text{sec}} \cdot \dfrac{60 \text{ sec}}{1 \text{ min}} \cdot \dfrac{60 \text{ min}}{1 \text{ hour}} \cdot \dfrac{1 \text{ mile}}{5280 \text{ ft}}$

   = 150 miles per hour

67. $\dfrac{4 \text{ sec}}{1} \cdot \dfrac{60 \text{ miles}}{\text{hour}} \cdot \dfrac{1 \text{ hour}}{60 \text{ min}} \cdot \dfrac{1 \text{ min}}{60 \text{ sec}} \cdot \dfrac{5280 \text{ ft}}{1 \text{ mile}}$

   = 352 feet, more than a city block

69. $\dfrac{25 \text{ hours}}{1 \text{ oil change}} \cdot \dfrac{100 \text{ feet}}{30 \text{ sec.}} \cdot \dfrac{1 \text{ mile}}{5280 \text{ ft}} \cdot \dfrac{3600 \text{ sec}}{1 \text{ hr}}$

   ≈ 56.8 miles per oil change

71. $\dfrac{30 \text{ min}}{1} \cdot \dfrac{6 \text{ miles}}{1 \text{ hour}} \cdot \dfrac{1 \text{ hour}}{60 \text{ min}} = 3$ miles

73. $\dfrac{250 \text{ mL}}{12 \text{ hours}} \cdot \dfrac{60 \text{ microdrops}}{1 \text{ mL}} \cdot \dfrac{1 \text{ hr}}{60 \text{ min}}$

   = 20.8 microdrops per minute

75. $\dfrac{40 \text{ miles}}{1} \cdot \dfrac{1 \text{ hr}}{300 \text{ miles}} \cdot \dfrac{60 \text{ min}}{1 \text{ hr}} = 8$ min.

77. $\dfrac{7.0 \text{ g Na}}{1} \cdot \dfrac{1 \text{ mole Na}}{23.0 \text{ g Na}} \cdot \dfrac{2 \text{ moles NaCl}}{2 \text{ moles Na}}$

   $\cdot \dfrac{58.5 \text{ g NaCl}}{1 \text{ mole NaCl}} = 17.8$ g NaCl

Section 5.2

1. 3 : 1, unsafe, slips

3. 4.5 : 1, unsafe, tips

5. 4 : 1, safe

7. $\dfrac{6}{8} \overset{?}{=} \dfrac{15}{20}$, $6(20) \overset{?}{=} 8(15)$, $120 = 120$

proportion

9. $\dfrac{4}{6} \overset{?}{=} \dfrac{6}{9}$, $4(9) \overset{?}{=} 6(6)$, $36 = 36$

proportion

11. $\dfrac{9}{21} \overset{?}{=} \dfrac{21}{35}$, $9(35) \overset{?}{=} 21(21)$, $315 \neq 441$

false statement

13. $\dfrac{2}{4} = \dfrac{5}{10}$, $2(10) = 4(5)$, $20 = 20$

$\dfrac{4}{2} = \dfrac{10}{5}$, $4(5) = 2(10)$, $20 = 20$

$\dfrac{10}{4} = \dfrac{5}{2}$, $10(2) = 4(5)$, $20 = 20$

$\dfrac{4}{10} = \dfrac{2}{5}$, $4(5) = 10(2)$, $20 = 20$

15. $\dfrac{3}{4} = \dfrac{x}{15}$, $4x = 15 \cdot 3$, $x = \dfrac{15 \cdot 3}{4}$,

$x = 11\frac{1}{4}$

17. $\dfrac{x}{5} = \dfrac{2}{3}$, $3x = 2 \cdot 5$, $x = \dfrac{5 \cdot 2}{3}$,

$x = 3\frac{1}{3}$

19. $\dfrac{7}{3} = \dfrac{5}{x}$, $7x = 3 \cdot 5$, $x = \dfrac{3 \cdot 5}{7}$,

$x = 2\frac{1}{7}$

21. $\dfrac{4}{x} = \dfrac{3}{7}$, $3x = 7 \cdot 4$, $x = \dfrac{7 \cdot 4}{3}$,

$x = 9\frac{1}{3}$

23. $0.45x = 36$, $x = \dfrac{36}{0.45}$, $x = 80$

or $\dfrac{45}{100} = \dfrac{36}{x}$, $45x = 36(100)$,

$x = \dfrac{3600}{45}$, $x = 80$

25. $56 = \dfrac{x}{100} \cdot 84$, $x = \dfrac{5600}{84}$, $x \approx 66.7\%$

or $\dfrac{56}{84} = \dfrac{x}{100}$, $84x = 56(100)$,

$x = \dfrac{5600}{84}$, $x \approx 66.7\%$

27. $\dfrac{60}{100} = \dfrac{x}{42}$, $100x = 60(42)$,

$x = \dfrac{2520}{100}$, $x = 25.2$

29. $\dfrac{x}{100} = \dfrac{56}{40}$, $40x = 5600$,

$x = \dfrac{5600}{40}$, $x = 140\%$

Section 5.2 (con't)

31. $\dfrac{18}{100} = \dfrac{x}{25}$, $100x = 450$,

$x = \dfrac{450}{100}$, $x = 4.5$

33. $0.64x = 16$, $x = \dfrac{16}{0.64}$, $x = 25$

or $\dfrac{64}{100} = \dfrac{16}{x}$, $64x = 16(100)$,

$x = \dfrac{1600}{64}$, $x = 25$

35. $\dfrac{x}{100} = \dfrac{104}{80}$, $80x = 104(100)$, $x = \dfrac{10400}{80}$,

$x = 130\%$

37. $\dfrac{x}{100} = \dfrac{28}{40}$, $40x = 28(100)$, $x = \dfrac{2800}{40}$,

$x = 70$

39. $\dfrac{x-1}{2} = \dfrac{x+3}{3}$, $3(x-1) = 2(x+3)$,

$3x - 3 = 2x + 6$, $x - 3 = 6$, $x = 9$

41. $\dfrac{x+4}{15} = \dfrac{x+2}{10}$, $10(x+4) = 15(x+2)$,

$10x + 40 = 15x + 30$, $40 = 5x + 30$,
$10 = 5x$, $x = 2$

43. $\dfrac{x+7}{9} = \dfrac{x+3}{6}$, $6(x+7) = 9(x+3)$,

$6x + 42 = 9x + 27$, $42 = 3x + 27$,
$15 = 3x$, $x = 5$

45. $\dfrac{3x}{2} = \dfrac{6x+3}{5}$, $15x = 2(6x+3)$,

$15x = 12x + 6$, $3x = 6$, $x = 2$

47. $\dfrac{2x-3}{7} = \dfrac{x+2}{4}$, $4(2x-3) = 7(x+2)$,

$8x - 12 = 7x + 14$, $x - 12 = 14$, $x = 26$

49. $\dfrac{4x+2}{5} = \dfrac{x-1}{2}$, $2(4x+2) = 5(x-1)$,

$8x + 4 = 5x - 5$, $3x + 4 = -5$,
$3x = -9$, $x = -3$

51. $\dfrac{x+2}{2} = \dfrac{3x+1}{5}$, $5(x+2) = 2(3x+1)$,

$5x + 10 = 6x + 2$, $10 = x + 2$, $x = 8$

53. $\dfrac{x}{65\,\text{in}} = \dfrac{1\,\text{m}}{39.37\,\text{in}}$, $39.37x\,\text{in} = 65\,\text{m in}$,

$x = \dfrac{65\,\text{m in}}{39.37\,\text{in}}$, $x \approx 1.65\,\text{m}$

55. $\dfrac{x}{15\,\text{mi}} = \dfrac{1\,\text{km}}{0.621\,\text{mi}}$, $0.621x\,\text{mi} = 15\,\text{km mi}$,

$x = \dfrac{15\,\text{km mi}}{0.621\,\text{mi}}$, $x \approx 24.15\,\text{km}$

57. $640\,\text{acres} \cdot \dfrac{1\,\text{mi}^2}{640\,\text{acres}} \cdot \dfrac{(5280\,\text{ft})^2}{1\,\text{mi}^2}$

$= 27{,}878{,}400\,\text{ft}^2$

59. $\dfrac{x}{6.5\,\text{ft}} = \dfrac{4}{1}$, $x = 4(6.5\,\text{ft})$, $x = 26\,\text{ft}$

Section 5.2 (con't)

61. $54 \text{ in} = 4.5 \text{ ft}, \quad \dfrac{4.5 \text{ ft}}{x} = \dfrac{1}{8}, \quad x = 36 \text{ ft}$

63. $\dfrac{8 \text{ ft}}{x} = \dfrac{7}{11}, \quad 7x = 88 \text{ ft}, \quad x = \dfrac{88 \text{ ft}}{7},$

$x \approx 12.57 \text{ ft}$

65. $\dfrac{13 \text{ miles}}{x} = \dfrac{1 \text{ mile}}{5280 \text{ ft}}, \quad x = 68,640 \text{ ft}$

$\text{slope} = \dfrac{\text{rise}}{\text{run}} = \dfrac{10 \text{ ft}}{68640 \text{ ft}} = \dfrac{1}{6864}$

However, unit analysis maybe more

sensible: $\dfrac{10 \text{ ft}}{13 \text{ mi}} \cdot \dfrac{1 \text{ mi}}{5280 \text{ ft}} = \dfrac{1}{6864}$

67. $\dfrac{x}{40 \text{ lb}} = \dfrac{500,000 \text{ units}}{150 \text{ lb}},$

$150x \text{ lb} = 20,000,000 \text{ units lb},$

$x \approx 130,000 \text{ units}$

69. $\dfrac{62.5}{100} = \dfrac{x}{3000}, \quad 100x = 187,500$

$x = 1875 \text{ ft}$

71. $\dfrac{9}{100} = \dfrac{x}{2}, \quad 100x = 18, \quad x = \dfrac{18}{100} \text{ mi}$

$0.18 \text{ mi} \cdot \dfrac{5280 \text{ ft}}{1 \text{ mi}} \approx 950 \text{ ft}$

73. $\dfrac{75 \text{ tagged}}{\text{x}} = \dfrac{10 \text{ tagged}}{65 \text{ total birds}},$

$10\text{x} = 4875, \quad \text{x} \approx 487 \text{ to } 488 \text{ total birds}$

75. Units are the same on both sides after cross multiplication.

Section 5.3

1. $\dfrac{7.5}{12} = \dfrac{h}{8}$, $12h = 7.5(8)$, $h = \dfrac{7.5 \cdot 8}{12}$,

$h = 5$ cm

3. Trace the rectangle. Draw a diagonal. Increase the base to 2.5 inches. Draw a vertical line at 2.5 inches up to diagonal. Vertical line is 1.25 inches.

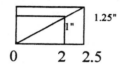

5. Trace the rectangle. Draw a diagonal. Decrease the base to 1.0 inches. Draw a vertical line at 1.0 inches up to diagonal. Vertical line is 0.2 inches.

7. Ratios are $\dfrac{16}{25} = \dfrac{16}{25}$, similar triangles

Corresponding sides:

Sides: AB and ND, AT and NE,

BT and DE

9. Ratios are $\dfrac{10}{29}$ and $\dfrac{29}{34}$, $\dfrac{10}{29} \neq \dfrac{29}{34}$,

triangles are not similar

11. Ratios are $\dfrac{6}{12}$ and $\dfrac{16}{32}$, both equal $\dfrac{1}{2}$,

figures are similar.

Corresponding sides are:

FG and HK, FR and HA,

RO and AW, GO and KW

13. Ratios are $\dfrac{11}{8}$ and $\dfrac{22}{16}$, $\dfrac{11}{8} = \dfrac{22}{16}$,

figures are similar.

Corresponding segments are:

radii RT and OE, diameters EN and DV

15. $\dfrac{28}{56} = \dfrac{n}{63}$, $56n = 28(63)$, $n = \dfrac{28(63)}{56}$,

$n = 31.5$

17. $\dfrac{12}{6} = \dfrac{n}{24}$, $6n = 12(24)$, $n = \dfrac{12(24)}{6}$,

$n = 48$

19. $\dfrac{35}{30} = \dfrac{s}{4}$, $30s = 35(4)$ $s = \dfrac{35(4)}{30}$,

$s \approx 4.7$ ft

21. $\dfrac{7}{1.4} = \dfrac{t}{4}$, $1.4t = 7(4)$, $t = \dfrac{7(4)}{1.4}$, $t = 20$ ft

23. $\dfrac{14}{27} = \dfrac{t}{78}$, $27t = 14(78)$, $t = \dfrac{14(78)}{27}$

$t \approx 40.4$ ft

Section 5.3 (con't)

25. $\dfrac{8}{5} = \dfrac{4+x}{x}$, $8x = 5(4+x) = 20+5x$

$3x = 20$, $x = \frac{20}{3}$, $x = 6\frac{2}{3}$

27. $\dfrac{15-5}{15} = \dfrac{6}{x}$, $10x = 15(6)$,

$10x = 90$, $x = 9$

29. $\dfrac{4}{3} = \dfrac{12}{12-x}$, $4(12-x) = 3(12)$,

$48 - 4x = 36$, $4x = 12$, $x = 3$

31. Let x = distance between origin and A. Using similar triangles:

$\dfrac{3}{x} = \dfrac{2}{4}$, $2x = 12$, $x = 6$, A = (6, 0)

B = (3, 2), from points given on graph

33. A = (0, 2), from points given on graph

Let y = distance from origin to B. The distance from A to B is then y - 2.

Using similar triangles:

$\dfrac{6}{3.5} = \dfrac{y}{y-2}$, $3.5y = 6(y-2)$

$3.5y = 6y - 12$, $2.5y = 12$, $y = 4.8$

B = (0, 4.8)

35. Think of the parking meter and its shadow as the height and base of a small triangle inside of a larger triangle. The larger triangle has a base of the distance from the street light (x) plus the shadow and a height of the street light.

$\dfrac{5}{4} = \dfrac{15}{x+4}$, $5(x+4) = 60$, $5x+20 = 60$

$5x = 40$, $x = 8$ ft

37. AB = 10 - x

39. BC = x - 10

41. Squares, circles and spheres are always similar.

Mid-Chapter 5 Test

1. 12xy to $15y^2z = 3(4)xy$ to $3(5)yyz =$

4x to 5yz

2. $\dfrac{(x-2)(x+2)}{x(x+2)} = \dfrac{x-2}{x}$

3. 18 inches = 1.5 feet

ratio is 1.5 : 6 = 1 : 4

4. 3 meters = 300 centimeters

ratio is 300 to 75 = 4 to 1

5. $\dfrac{16}{100} = \dfrac{11.2}{x}$, $16x = 11.2(100)$, $x = 70$

or $0.16x = 11.2$, $x = \dfrac{11.2}{0.16}$, x = 70

6. $\dfrac{360}{x} = \dfrac{250}{100}$, $250x = 360(100)$, $x = 144$

or $2.50x = 360$, $x = \dfrac{360}{2.50}$, x = 144

7. $\dfrac{\text{new - original}}{\text{original}} = \dfrac{30,000 - 24,000}{24,000}$

$= \dfrac{6,000}{24,000} = \dfrac{1}{4} = 0.25$, 25 % increase

8. $\dfrac{\text{new - original}}{\text{original}} = \dfrac{13.99 - 15.99}{15.99}$

$= \dfrac{-2}{15.99} \approx -0.125$, 12.5% decrease

9. One possible answer is the fuel usage in an aircraft.

10. One possible answer is production numbers in a chocolate factory.

11. $\dfrac{5}{x} = \dfrac{500,000}{1}$, $x = \dfrac{5}{500,000}$, $x = \dfrac{1}{100,000}$

12. $\dfrac{3}{5} = \dfrac{16}{x}$, $3x = 5(16)$, $x = \dfrac{80}{3}$, $x = 26\frac{2}{3}$

13. $\dfrac{x}{12} = \dfrac{10}{8}$, $8x = 120$, $x = \dfrac{120}{8}$, $x = 15$

14. $\dfrac{x+5}{8} = \dfrac{2x-1}{12}$, $12(x+5) = 8(2x-1)$

$12x + 60 = 16x - 8$, $68 = 4x$, $x = \dfrac{68}{4}$, $x = 17$

15. $\dfrac{5x-2}{3x} = \dfrac{4}{3}$, $3(5x-2) = 4(3x)$

$15x - 6 = 12x$, $3x = 6$, $x = 2$

16. $\dfrac{a}{b} = \dfrac{c}{d}$, $bc = ad$, $b = \dfrac{ad}{c}$

17. $\dfrac{6}{8} = \dfrac{x}{15}$, $8x = 6(15)$, $x = \dfrac{6(15)}{8}$, $x = 11\frac{1}{4}$

18. A = (5, 0), from the points in the figure

Let y = the height between B and (7, 0).

Using similar triangles:

$\dfrac{5}{7} = \dfrac{4}{y}$, $5y = 4 \cdot 7$, $y = \dfrac{28}{5}$, $y = 5.6$

B = (7, 5.6)

Mid-Chapter 5 Test (con't)

19. $2 + 3 + 5 = 10$ total parts,

$1500 \div 10 = 150$ per part

fat $= 2(150) = 300$

carbohydrate $= 3(150) = 450$

protein $= 5(150) = 750$

or let x $=$ one part, $2x + 3x + 5x = 1500$,

$10x = 1500$, $x = 150$,

$2x = 2(150) = 300$,

$5x = 5(150) = 750$

20. $1 + 1 + 2 = 4$ parts

$180 \div 4 = 45$ per part

Angles are $45°$, $45°$, and $2(45°) = 90°$

or let x $= 1$ part, $1x + 1x + 2x = 180$,

$4x = 180$, $x = 45$,

$2x = 2(45) = 90$

Angles are $45°$, $45°$ and $90°$

21. $\dfrac{1}{12} = \dfrac{48}{x}$, $x = 48(12)$, $x = 576$ inches

$576 \text{ inches} \cdot \dfrac{1 \text{ ft}}{12 \text{ inches}} = 48 \text{ ft}$

22. $\dfrac{\text{clipped fin}}{\text{not clipped}} = \dfrac{\text{clipped fin}}{\text{not clipped}}$,

$\dfrac{10,000}{n} = \dfrac{250}{10}$, $250n = 100,000$

$n = \dfrac{100,000}{250} = 400$, assumes all

hatchery fish survived and the hatchery

fish mixed completely with native fish.

23. a. $0.059(5,000,000,000) =$

$\$295,000,000$ annual interest.

b. $\dfrac{295,000,000}{1 \text{ yr}} \cdot \dfrac{1 \text{ yr}}{365 \text{ days}} \cdot \dfrac{1 \text{ day}}{24 \text{ hr}}$

$\cdot \dfrac{1 \text{ hr}}{60 \text{ min}} \cdot \dfrac{1 \text{ min}}{60 \text{ sec}} \approx \$9.35$ per second

Newspaper over-stated the interest by a

small amount - or rounded to the nearest

ten cents.

Section 5.4

1. **a.** $57 + 57 + 57 + 58 + 61 = 290$

   $290 \div 5 = 58$, mean $= 58$

   median $= 57$, mode $= 57$

   b. $51 + 54 + 57 + 61 + 68 = 291$

   $291 \div 5 = 58.2$, mean $= 58.2$

   median $= 57$, no mode

   c. $46 + 52 + 61 + 64 + 69 = 292$

   $292 \div 5 = 58.4$, mean $= 58.4$

   median $= 61$, no mode

   d. $46 + 52 + 54 + 56 + 65 = 273$

   $273 \div 5 = 54.6$, mean $= 54.6$

   median $= 54$, no mode

3. **a.** $315 + 340 + 350 + 365 + 365 + 375$
   $+ 385 + 395 + 395 + 410 + 415 + 415 +$
   $430 + 440 = 5395$, $5395 \div 14 \approx 385.36$

   mean $= 385.4$,

   median $= (385 + 395) \div 2 = 390$

   3 modes: 365, 395 and 415

   b. $285 + 325 + 375 + 400 + 405 + 410$
   $+ 410 + 435 + 435 + 510 + 515 + 515 +$
   $550 + 550 = 6120$, $6120 \div 14 \approx 437.14$

   mean $= 437.1$,

   median $= (410 + 435) \div 2 = 422.5$

   4 modes: 410, 435, 515, and 550

5. No; for example the mean of 5 and 7 is 6 and equals the mean of 4 and 8.

7. One low or high grade will move the mean away from the median.

9. Is not; for example the median of 2, 3, 5 $= 3 =$ median of 0, 3, 25.

11. When we want to reduce the effect of large or small data.

13. No, median means half above.

15. Some data are comparatively very small.

17. There would not be time to test for everything.

19. It is in the *middle* of the traffic lanes.

21. GRCC $= 4(5) + 3(3) + 2(1) = 31$

    LCC $= 5(5) + 0(3) + 4(1) = 29$

    BCC $= 1(5) + 7(3) + 4(1) = 30$

23. $0.85 + 0.70 + 0.80 + 2(0.95) = 4.25$

    $4.25 \div 5 = 0.85$

25. $0.80 + 0.80 + 0.70 + 2(0.60) = 3.50$

    $3.50 \div 5 = 0.70$

27. $\dfrac{0.76 + 0.81 + 0.72 + 2(x)}{5} = 0.80$

    $2.29 + 2x = 4$, $2x = 1.71$, $x \approx 0.86$

Section 5.4 (con't)

29. $\dfrac{0.50 + 0.70 + 0.70 + 2(x)}{5} = 0.80$

$1.9 + 2x = 4,\ 2x = 2.1,\ x = 1.05$

Student needs over 100 on final, not possible.

~~midpoint~~

31. **a.** $\left(\dfrac{0+5}{2}, \dfrac{4+2}{2}\right) = \left(\dfrac{5}{2}, \dfrac{6}{2}\right) = (2.5, 3)$

   b. $\left(\dfrac{-1+3}{2}, \dfrac{3+(-3)}{2}\right) = \left(\dfrac{2}{2}, \dfrac{0}{2}\right) = (1, 0)$

33. **a.** $\left(\dfrac{a+0}{2}, \dfrac{0+b}{2}\right) = \left(\dfrac{a}{2}, \dfrac{b}{2}\right)$

   b. $\left(\dfrac{a+0}{2}, \dfrac{a+0}{2}\right) = \left(\dfrac{a}{2}, \dfrac{a}{2}\right)$

   c. $\left(\dfrac{0-0}{2}, \dfrac{b+0}{2}\right) = \left(0, \dfrac{b}{2}\right)$

35. Between $(0, 0)$ and $(0, 6)$ the midpoint

   is: $\left(0, \dfrac{6}{2}\right) = (0, 3)$

   Between $(0, 0)$ and $(9, 0)$ the midpoint

   is: $\left(\dfrac{9}{2}, 0\right) = (4.5, 0)$

   Between $(0, 6)$ and $(9, 0)$ the midpoint

   is: $\left(\dfrac{9}{2}, \dfrac{6}{2}\right) = (4.5, 3)$

   The centroid is at:

   $\left(\dfrac{0+0+9}{3}, \dfrac{0+6+0}{3}\right) = (3, 2)$

37. Between $(0, 0)$ and $(0, 6)$ the midpoint

   is: $\left(0, \dfrac{6}{2}\right) = (0, 3)$

   Between $(0, 0)$ and $(8, 0)$ the midpoint

   is: $\left(\dfrac{8}{2}, 0\right) = (4, 0)$

   Between $(0, 6)$ and $(8, 0)$ the midpoint

   is: $\left(\dfrac{8}{2}, \dfrac{6}{2}\right) = (4, 3)$

   The centroid is at:

   $\left(\dfrac{0+0+8}{3}, \dfrac{0+6+0}{3}\right) = (2\tfrac{2}{3}, 2)$

Section 5.5

1. Value

3. Quantity

5. Value

7. Value

9. Value

11. a.

Quantity	Value	Q•V
5 kg Peanuts	$8.80/kg	5(8.80) = $44.00
2 kg Cashews	$24.20/kg	2(24.20) = $48.40
7 kg		$92.40

b. The sum of the quantity column is the total kg of peanuts and cashews.

c. $92.40 ÷ 7 kg = $13.20 per kg

d. The sum of the Q•V column is the total worth of peanuts and cashews.

13. a.

Quantity	Value	Q•V
3 lbs Grapes	$0.98/lb	3(0.98) = $2.94
5 lbs Potatoes	$0.49/lb	5(0.49) = $2.45
2 lbs Broccoli	$0.89/lb	2(0.89) = $1.78
10 lbs		$7.17

b. The sum of the quantity column is the total pounds of produce purchased.

c. $7.17 ÷ 10 lbs ≈ $0.72 per pound

d. The sum of the Q•V column is the total amount of the purchase.

15. a.

Quantity	Value	Q•V
15 dimes	$0.10	15(0.10) = $1.50
20 quarters	$0.25	20(0.25) = $5.00
35		$6.50

b. The sum of the quantity column is the total number of coins.

c. $6.50 ÷ 35 ≈ $0.19

d. The sum of the Q•V column is the total dollar amount.

Section 5.5 (con't)

17. a.

Quantity	Value	Q•V
$1500	0.09	1500(0.09) = $135
$1500	0.06	1500(0.06) = $90
$3000		$225

b. The sum of the quantity column is the total amount invested.

c. $6.50 ÷ $3000 = 0.075, 7.5% average rate

d. The sum of the Q•V column is the total return on investment.

19. a.

Quantity	Value	Q•V
100 lb	0.12	100(0.12) = 12.0
50 lb	0.15	50(0.15) = 7.5
150 lb		19.5

b. The sum of the quantity column is the total pounds of dog food.

c. 19.5 ÷ 150 = 0.13, 13% protein

d. The sum of the Q•V column is the total pounds of protein.

21. a.

Quantity	Value	Q•V
150 lb	0.10	150(0.10) = 15
25 lb	0	25(0) = 0
175 lb		15

b. The sum of the quantity column is the total pounds of dried grasses.

c. 15 ÷ 175 ≈ 0.086, 8.6% protein

d. The sum of the Q•V column is the total pounds of protein.

23. a.

Quantity	Value	Q•V
5 hrs D	1	5(1) = 5
4 hrs C	2	4(2) = 8
3 hrs B	3	3(3) = 9
12 hrs		22

b. The sum of the quantity column is the total hours.

c. 22 ÷ 12 hrs ≈ 1.83

d. The sum of the Q•V column is the total points.

Section 5.5 (con't)

25. a.

Quantity	Value	Q•V
3 hr	80 kph	3(80) = 240 km
2 hr	30 kph	2(30) = 60 km
5 hr		300 km

b. The sum of the quantity column is the total hours.

c. 300 km ÷ 5 hrs = 60 kph

d. The sum of the Q•V column is the total km.

27. a.

Quantity	Value	Q•V
150 ml	18	150(18) = 2700
100 ml	3	100(3) = 300
250 ml		3000

b. The sum of the quantity column is the total ml.

c. 3000 ÷ 250 ml = 12

d. The sum of the Q•V column is the total moles of sulfuric acid.

29. $300 - x$, $x + (300 - x) = 300$

31. $\$15,000 - x$, $x + (\$15,000 - x) = \$15,000$

33. $16 - x$, $x + (16 - x) = 16$

35.

Quantity	Value	Q•V
200 Boeing	$54	200($54) = $10,800
x Nike	$71	$71x
200 + x		$25,000

$\$10,800 + \$71x = \$25,000$

$71x = 14,200$, $x = 200$ shares of Nike

37.

Quantity	Value	Q•V
30 hrs	$5.80	30($5.80) = $174.00
x hrs	$7.20	$7.20x
30 + x	$6.36	$174 + $7.20x

$\$6.36(30 + x) = \$174 + \$7.20x$

$190.80 + 6.36x = 174 + 7.20x$

$16.80 = 0.84x$, $x = 20$ hours

39. From exercise #37 with $7 average wage: $7(30 + x) = \$174 + \$7.20x$

$210 + 7x = 174 + 7.20x$

$36 = 0.20x$, $x = 180$ hours, not reasonable.

Chapter 5   **123**

Section 5.5 (con't)

41.

Quantity	Value	Q•V
300 - x Colombian	$7.25	(300 - x)$7.25
x Sumatran	$10	$10x
300	$8.35	300($8.35) = $2505

$7.25(300 - x) + 10x = $2505

2175 - 7.25x + 10x = 2505

2.75x = 330, x = 120

300 - 120 = 180 lbs Colombian

120 lbs Sumatran

43 & 45 use the table below

Quantity	Value	Q•V
x peanuts	$10	$10x
50 - x cashews	$24	24(50 - x) = 1200 - 24x
50		1200 - 14x

43. Place $12 as average value (below $24),

50($12) = $1200 - $14x, solve for x.

600 = 1200 - 14x, 14x = 600

x ≈ 42.9 kg peanuts

50 - 42.9 ≈ 7.1 kg cashews

45. Place $18 as average value (below $24),

50($18) = $1200 - $14x, solve for x.

900 = 1200 - 14x, 14x = 300

x ≈ 21.4 kg peanuts

50 - 21.4 ≈ 28.6 kg cashews

47.

Quantity	Value	Q•V
A	0.05	0.05A
B or (15,000 - A)	0.08	0.08(15,000 - A) = 1200 - 0.08A
$15,000		1200 - 0.03A

a. A = 0, B = $15,000

Earnings = $15,000(0.08) = $1200

Average rate = 0.08 = 8%

b. A = $7,500, B = $7,500

Earnings = $7,500(0.05) + $7,500(0.08) = $375 + $600 = $975

Average rate = 975 ÷ 15,000 = 0.065 = 6.5%

Section 5.5 (con't)

49. Using the table from exercise #47

The sum of Q•V equals the interest earned:

a. $1200 - 0.03A = $1060, solve for A

$140 = 0.03A$, $A \approx 4666.67

$B \approx $15,000 - $4666.67 \approx $10,333.33$

b. $1200 - 0.03A = $825

$375 = 0.03A$, $A = $12,500$

$B = $15,000 - $12,500 = $2,500$

51.

Quantity	Value	Q•V
20	1.85	37
x	3	3x
20 + x	2.00	37 + 3x

$2(20 + x) = 37 + 3x$

$40 + 2x = 37 + 3x$

$x = 3$ hours

53.

Quantity	Value	Q•V
12	140°	1680
x	60°	60x
12 + x	103°	1680 + 60x

$103(12 + x) = 1680 + 60x$

$1236 + 103x = 1680 + 60x$

$43x = 444$, $x \approx 10.3$ gal

55.

Quantity	Value	Q•V
x	140°	140x
15	35°	525
15 + x	103°	525 + 140x

$103(15 + x) = 525 + 140x$

$1545 + 103x = 525 + 140x$

$1020 = 37x$, $x \approx 27.6$ gal

57.

Quantity	Value	Q•V
x	140°	140x
x	35°	35x
2x	t	175x

$2x(t) = 175x$, $t = \dfrac{175x}{2x}$, $t = 87.5°$

87.5° would be the result for any equal quantities because the quantity x drops out of the equation.

59.

Quantity	Value	Q•V
x ml	3	3x
300 ml	18	5400
x + 300	12	3x + 5400

$12(x + 300) = 3x + 5400$

$12x + 3600 = 3x + 5400$

$9x = 1800$, $x = 200$ ml

Section 5.5 (con't)

61. In example 5 we multiply money invested by interest rate to find interest earned. In example 7 we multiply the number of shares by the price per share to get the money invested.

63. The average value is closer to the value of the item with the larger quantity.

65. If the quantity value is very large compared to others it will have a larger effect.

67. The greatest percent decrease is from $0.20 to $0.05. The smallest percent decrease was from $0.30 to $0.20.

$$\frac{0.50 - 1.00}{1.00} = -0.50, \; 50\% \text{ decrease}$$

$$\frac{0.30 - 0.50}{0.50} = -0.40, \; 40\% \text{ decrease}$$

$$\frac{0.20 - 0.30}{0.30} \approx -0.333, \; 33.3\% \text{ decrease}$$

$$\frac{0.05 - 0.20}{0.20} = 0.75, \; 75\% \text{ decrease}$$

Chapter 5 Review

1. $16x^2$ to $4x^4 = 4x^2(4)$ to $4x^2(x^2) =$

  4 to x^2

3. 5 feet = 60 inches

  ratio is $60 : 18 = 6(10) : 6(3) = 10 : 3$

5. $0.75x = 108, \ x = 144$

7. Answers will vary.

9. Let $x = 1$ share, $5x + 5x + 1x = 121,000,$

  $11x = 121,000,$

  $x = 11,000$ cases per share

  Pepsi $= 5(11,000) = 55,000$

  Coca Cola $= 5(11,000) = 55,000$

  Other $= 1(11,000) = 11,000$

11. $0.30(1500) = 450$ cal,

  $\dfrac{450\,\text{cal}}{1} \cdot \dfrac{1\,\text{g fat}}{9\,\text{cal}} = 50\,\text{g}$

13. $\dfrac{\$42.49 - \$49.99}{\$49.99} = \dfrac{-7.50}{49.99} \approx -0.15$

  15% decrease

15. $\dfrac{\$0.60\,\text{Canadian}}{1\,\text{liter}} \cdot \dfrac{\$0.70\,\text{U.S.}}{\$1\,\text{Canadian}}$

  $\cdot \dfrac{1\,\text{liter}}{1.0567\,\text{quart}} \cdot \dfrac{4\,\text{quarts}}{1\,\text{gallon}}$

  $\approx \$1.59$ U.S. per gal

17. $\dfrac{1}{8} = \dfrac{30}{x}, \ x = 240$ in.

  $240\,\text{in} \cdot \dfrac{1\,\text{ft}}{12\,\text{in}} = 20\,\text{ft}$

19. $\dfrac{3}{1} = \dfrac{x}{1.5}, \ x = 4.5$ gal

21. $\dfrac{2}{6} = \dfrac{3}{x}, \ 2x = 18, \ x = 9$

23. $\dfrac{x+1}{2} = \dfrac{4x-1}{6}, \ 6(x+1) = 2(4x-1)$

  $6x + 6 = 8x - 2, \ 2x = 8, \ x = 4$

25. $\dfrac{a}{b} = \dfrac{c}{d}, \ ad = bc, \ d = \dfrac{bc}{a}$

27. a. $\dfrac{3}{5} = \dfrac{x}{14}, \ x = \dfrac{3(14)}{5}, \ x = 8.4$

  b. $\dfrac{5}{7} = \dfrac{14-x}{14}, \ 7(14-x) = 70$

  $98 - 7x = 70, \ 7x = 28, \ x = 4$

  c. $\dfrac{x}{h} = \dfrac{3}{b}, \ x = \dfrac{3h}{b}$

  d. $\dfrac{x}{w} = \dfrac{l-4}{l}, \ x = \dfrac{w(l-4)}{l}$

Chapter 5 Review (con't)

29. The slope of the line is:

$$m = \frac{4-3}{5-3} = \frac{1}{2}$$

Solving for the y-intercept using (3, 3):

$$b = 3 - \tfrac{1}{2}(3) = \tfrac{3}{2} = 1.5$$

A = (0, 1.5)

Solving for the x-intercept:

$$0 = \tfrac{1}{2}x + \tfrac{3}{2}, \; 0 = x + 3, \; x = -3$$

B = (-3, 0)

31.

Quantity	Value	Q•V
$10,000	0.08	$10,000(0.08) = $800
$5000	0.035	$5000(0.035) = $175
$15,000	975 ÷ 15,000 = 0.065	$975

33.

Quantity	Value	Q•V
8 liters	90° C	8(90) = 720
50 liters	5°	50(5) = 250
58 liters	970 ÷ 58 ≈ 16.7°	970

35.

Quantity	Value	Q•V
12 cr A	4 points	12(4) = 48 cr pt
4 cr B	3 points	4(3) = 12 cr pt
16 cr	60 ÷ 16 = 3.75 points	60 cr pt

37.

Quantity	Value	Q•V
$10,000	0.03	$10,000(0.03) = $300
x	0.08	0.08x
$10,000 + x	0.06	0.08x + 300

$$0.06(\$10,000 + x) = 0.08x + \$300$$

$$\$600 + 0.06x = 0.08x + \$300$$

$$0.02x = \$300, \; x = \$15,000$$

39. 31.6 + 31.7 + 31.8 + 31.8 = 126.9

126.9 ÷ 4 ≈ 31.73, Mean = 31.73

$$\text{Median} = \frac{31.7 + 31.8}{2} = 31.75$$

Mode = 31.8

41. 44.0 + 44.5 + 45.0 + 45.0 + 45.5 = 224

224 ÷ 5 = 44.8, Mean = 44.8

Median = 45.0, Mode = 45.0

Chapter 5 Review (con't)

43. a. Midpoints:

$$EG = \left(\frac{5+(-3)}{2}, \frac{12+2}{2}\right) = (1, 7)$$

$$EF = \left(\frac{5+(-3)}{2}, \frac{2+2}{2}\right) = (1, 2)$$

$$HE = \left(\frac{-3+(-3)}{2}, \frac{12+2}{2}\right) = (-3, 7)$$

$$HF = \left(\frac{-3+5}{2}, \frac{12+2}{2}\right) = (1, 7)$$

b. Slopes:

$$HF = \frac{12-2}{-3-5} = \frac{10}{-8} = -\frac{5}{4}$$

$$EG = \frac{12-2}{5-(-3)} = \frac{10}{8} = \frac{5}{4}$$

c. Slope of HE $= \dfrac{12-2}{-3-(-3)} = \dfrac{10}{0}$

Division by zero is undefined

d. Centroid:

$$\left(\frac{-3+5+5+(-3)}{4}, \frac{12+12+2+2}{4}\right)$$

$$= (1, 7)$$

45. Centroid =

$$\left(\frac{-1+1+3}{3}, \frac{-2+7+-2}{3}\right) = (1, 1)$$

47. $\dfrac{2(100)+2(85)+1(95)+1(75)}{6}$

$$= \frac{540}{6} = 90$$

Chapter 5 Test

1. & 2.   Answers will vary

3. $\dfrac{7ab^3}{28a^2b^2} = \dfrac{b}{4a}$

4. $\dfrac{a+b}{(a+b)(a-b)} = \dfrac{1}{a-b}$

5. $\dfrac{10\,\text{ft}}{24\,\text{in}} \cdot \dfrac{12\,\text{in}}{1\,\text{ft}} = \dfrac{5}{1}$

6. $1.25(x) = 105,\ x = 84$

7. $\dfrac{2\,\text{packs}}{1\,\text{day}} \cdot \dfrac{1\,\text{carton}}{20\,\text{packs}} \cdot \dfrac{\$35}{1\,\text{carton}} \cdot \dfrac{365\,\text{days}}{1\,\text{yr}} =$

$1,277.50$ per year

8. $\dfrac{3.5}{5} = \dfrac{x}{2},\ x = \dfrac{7}{5},\ x = 1.4\ \text{ft}$

9. $\dfrac{2}{5} = \dfrac{13}{x},\ 2x = 65,\ x = 32.5$

10. $\dfrac{x+1}{4} = \dfrac{x-3}{3},\ 3(x+1) = 4(x-3)$

$3x + 3 = 4x - 12,\ x = 15$

11. $\dfrac{0.85 + 0.91 + x}{3} = 0.90$

$1.76 + x = 2.70,\ x = 0.94\ = 94\%$

12. Median = $0.29

$0.29 + \$0.29 + \$0.29 + \$0.29 + \$1.79 =$
$2.95,\ \$2.95 \div 5 = \$0.59 = \text{Mean}$

13. Median = $0.69

$0.19 + \$0.29 + \$0.69 + \$0.69 + \$1.09 =$
$2.95,\ \$2.95 \div 5 = \text{Mean} = \$0.59$

14. Answers will vary

15. Midpoints:

$\left(\dfrac{3+7}{2}, \dfrac{2+8}{2}\right) = (5,\,5)$

$\left(\dfrac{7+8}{2}, \dfrac{8+2}{2}\right) = (7.5,\,5)$

$\left(\dfrac{8+3}{2}, \dfrac{2+2}{2}\right) = (5.5,\,2)$

16. Centroid:

$\left(\dfrac{3+7+8}{3}, \dfrac{2+8+2}{3}\right) = (6,\,4)$

17.

Quantity	Value	Q•V
$17,000	0.058	$17,000(0.058) = \$986$
$2,000	0.149	$2,000(0.149) = \$298$
$19,000 (Total debt)	$1284 \div \$19,000 \approx 0.068$ (avg rate on total debt)	$1284 (Total interest on debt)

Chapter 5 Test (con't)

18. a. $\dfrac{12}{15} = \dfrac{x}{20}$, $x = \dfrac{240}{15}$, $x = 16$

b. $\dfrac{5}{9} = \dfrac{6}{x+6}$, $5(x+6) = 54$

$x + 6 = 10.8$, $x = 4.8$

19.

Quantity	Value	Q•V
60 credits	3.40	$60(3.40) = 204$
x credits	4.00	4.00x
x + 60	3.50	4.00x + 204

$3.50(x + 60) = 4.00x + 204$

$3.50x + 210 = 4.00x + 204$

$6 = 0.50x$, $x = 12$ credits

20. $\dfrac{120}{12} = \dfrac{100}{I_1}$, $I_1 = \dfrac{1200}{120}$, $I_1 = 10\,\text{mA}$

V and I are not proportional in this problem. The products from cross multiplication are different.

Section 6.1

1. **a.** $-9 + 4 = -5$

  b. $-8 + 3 = -5$

  c. $-5 - 8 = -5 + (-8) = -13$

  d. $-7 - 3 = -7 + (-3) = -10$

  e. $-9 - (-5) = -9 + 5 = -4$

  f. $3 - (-4) = 3 + 4 = 7$

3. **a.** $5a + 3b - 2c - 4a - 6c + 9b$

  $= 5a - 4a + 3b + 9b - 2c - 6c$

  $= a + 12b - 8c$, trinomial

  b. $6m + 2n - 6p - 3m - 3n + 6p$

  $= 6m - 3m + 2n - 3n - 6p + 6p = 3m - n$

  binomial

  c. $8y + 5y + 5y - 5y + 8y = 21y$

  monomial

  d. $(x^2 + 3x) - (4x + 12)$

  $= x^2 + 3x - 4x - 12 = x^2 - x - 12$

  trinomial

  e. $(x^2 - 2x + 3) - (2x^2 - 4x + 6)$

  $= x^2 - 2x + 3 - 2x^2 + 4x - 6$

  $= -x^2 + 2x - 3$, trinomial

5. $5 - 3x^2 + 5x - x^3$

  $= -x^3 - 3x^2 + 5x + 5$

7. $x^2 - x^4 + 1 - x = -x^4 + x^2 - x + 1$

9. **a.** $x^2 + 2x + 3x + 6 = x^2 + 5x + 6$

  trinomial

  b. $3x^2 + 6x + x + 2 = 3x^2 + 7x + 2$

  trinomial

  c. $5 - 4(x - 3) = 5 - 4x + 12 = -4x + 17$

  binomial

  d. $6x^2 + 2x + 3x + 1 = 6x^2 + 5x + 1$

  trinomial

  e. $a^2 - ab + ab - b^2 = a^2 - b^2$

  binomial

11. **a.** $2y(-x + 2y) + x(x - 2y)$

  $= -2xy + 4y^2 + x^2 - 2xy$

  $= x^2 - 4xy + 4y^2$

  trinomial

  b. $x^2 - 4y^2 - 2xy + 2xy = x^2 - 4y^2$

  binomial

  c. $x^3 + 2x^2 + x + x^2 + 2x + 1$

  $= x^3 + 3x^2 + 3x + 1$, 4 - term polynomial

  d. $x(4 + 4x + x^2) - 2(4 + 4x + x^2)$

  $= 4x + 4x^2 + x^3 - 8 - 8x - 2x^2$

  $= x^3 + 2x^2 - 4x - 8$, 4 - term polynomial

Section 6.1 (con't)

11. e. $b^3 - ab^2 + a^2b + a^3 - a^2b + ab^2$

$= a^3 + b^3$, binomial

13. Length = a + b, Width = 2b

$P = (a + b) + 2b + (a + b) + 2b,$

$P = 2a + 6b$

$A = (a + b)(2b), A = 2ab + 2b^2$

15. Length = x + 2, Width = x + 1

$P = 2(x + 2) + 2(x + 1),$

$P = 2x + 4 + 2x + 2, P = 4x + 6$

$A = (x + 2)(x + 1)$

Counting the tiles $A = x^2 + 3x + 2$

17. a. $P = 2(0.75a) + 2(1.75a),$

$P = 1.5a + 3.5a, P = 5a$

b. $P = 4x$

c. $P = (2x - 3) + x + 2x,$

$P = 2x - 3 + x + 2x, P = 5x - 3$

d. $P = 5(\frac{1}{2}x), P = \frac{5}{2}x, P = 2\frac{1}{2}x$

e. $P = 2(1.5) + 2(0.5\pi), P = 3 + 1\pi,$

$P \approx 6.14$

19. a. Let w = missing width;

$10a + 6c = 2(5a) + 2w,$

$10a + 6c = 10a + 2w, 6c = 2w, w = 3c$

$A = (5a)(3c), A = 15ac$

b. Let l = missing length;

$3x + 5y = 2(1.5x) + 2l,$

$3x + 5y = 3x + 2l, 5y = 2l, l = 2.5y$

$A = (1.5x)(2.5y), A = 3.75xy$

c. Let s = side of square;

$20y = 4s, s = 5y$

$A = (5y)(5y), A = 25y^2$

21. a. $2a(a + 2b) = 2a^2 + 4ab$

b. $2b(2a + b) = 4ab + 2b^2$

c. $x(x + 3) = x^2 + 3x$

d. $x(2x + 1) = 2x^2 + x$

23. a. $2x(x^2 + 3x) = 2x^3 + 6x^2$

b. $x^2(x - 1) = x^3 - x^2$

c. $x^2(x^2 + 2x + 1) = x^4 + 2x^3 + x^2$

d. $ab(b^2 - 1) = ab^3 - ab$

e. $b^2(a - b) = ab^2 - b^3$

f. $a^2(1 + b - b^2) = a^2 + a^2b - a^2b^2$

25. $5x - 3x(1 - x) = 5x - 3x + 3x^2$

$= 3x^2 + 2x$

Section 6.1 (con't)

27. $8b + 2b(b + 1) = 8b + 2b^2 + 2b$

$= 2b^2 + 10b$

29. $4b - 2b(5 - b) = 4b - 10b + 2b^2$,

$= 2b^2 - 6b$

31.

33.

35. a. $111 = (3)(37)$

  b. $91 = (7)(13)$

37. a. gcf = 9, $\dfrac{36}{99} = \dfrac{9 \cdot 4}{9 \cdot 11} = \dfrac{4}{11}$

  b. gcf = 66, $\dfrac{66}{990} = \dfrac{66 \cdot 1}{66 \cdot 15} = \dfrac{1}{15}$

  c. gcf = 37, $\dfrac{185}{999} = \dfrac{37 \cdot 5}{37 \cdot 27} = \dfrac{5}{27}$

  d. gcf = m, $\dfrac{mn}{mp} = \dfrac{n}{p}$

37. e. gcf = 4n, $\dfrac{4np}{24mn} = \dfrac{4n \cdot p}{4n \cdot 6m} = \dfrac{p}{6m}$

39.

Factor	y^2	$+2xy$	$+x^2$
y	y^3	$+2xy^2$	$+x^2y$

41.

Factor	a	-2b	$+3b^2$
2ab	$2a^2b$	$-4ab^2$	$+6ab^3$

43. gcf = x,

$x^3 + 4x^2 + 4x = x(x^2 + 4x + 4)$

45. gcf = b,

$a^2b + ab^2 + b^3 = b(a^2 + ab + b^2)$

47. gcf = 2x, $6x^2 + 2x = 2x(3x + 1)$

49. gcf = 3y, $15y^2 - 3y = 3y(5y - 1)$

51. gcf = 5xy,

$15x^2y + 10xy^2 = 5xy(3x + 2y)$

53. Using gcf = 4, $-4x - 12 = 4(-x - 3)$

Using gcf = -4, $-4x - 12 = -4(x + 3)$

55. Using gcf = 2y,

$-2xy + 4y^2 = 2y(-x + 2y)$

Using gcf = -2y,

$-2xy + 4y^2 = -2y(x - 2y)$

Section 6.1 (con't)

57. Using gcf = 4,

$-12x^2 - 8x - 8 = 4(-3x^2 - 2x - 2)$

Using gcf = -4,

$-12x^2 - 8x - 8 = -4(3x^2 + 2x + 2)$

59. Using gcf = y^2,

$-y^2 + 4y^3 - 8y^4 = y^2(-1 + 4y - 8y^2)$

Using gcf = $-y^2$,

$-y^2 + 4y^3 - 8y^4 = -y^2(1 - 4y + 8y^2)$

61. a.

	1 - 2b
2a	2a - 4ab

b.

	x + y
xy	$x^2y + xy^2$

63. 2 terms, x^2 and 5xy

65. 1 term, $\frac{1}{2}gt^2$

67. 3 terms, $4x^2$, 8x and 8

69. 3 factors, 4, (x + 2) and (x + 2)

71. 4 factors, 2, π, r and (h + r)

73. 4 factors, 2, x, y and z

75. Exercise 63 is exercise 70 after applying the distributive property.

77. The distributive property, a(b + c) = ab + ac, changes the product of two **factors** into the **sum** of two **terms**.

79. Variables and exponents are identical.

81. Find the largest factor that is in common to all terms.

Section 6.2

1. $(x + 1)(x + 1) = x^2 + 2x + 1$

3. $(a + 2b)(2a + b) = 2a^2 + 5ab + 2b^2$

5. $(a + 2b)(a + 2b) = a^2 + 4ab + 4b^2$

7.

Multiply	2x	+5
x	$x(2x) = 2x^2$	+5x
-4	$-4(2x) = -8x$	$-4(+5) = -20$

$(x - 4)(2x + 5) = 2x^2 + 5x - 8x - 20$

$= 2x^2 - 3x - 20$

9.

Multiply	x	-2
2x	$2x(x) = 2x^2$	$2x(-2) = -4x$
+3	+3x	$+3(-2) = -6$

$(2x + 3)(x - 2) = 2x^2 - 4x + 3x - 6$

$= 2x^2 - x - 6$

11.

Multiply	x	-2
2x	$2x(x) = 2x^2$	$2x(-2) = -4x$
-1	-x	$-1(-2) = +2$

$(2x - 1)(x - 2) = 2x^2 - 4x - x + 2$

$= 2x^2 - 5x + 2$

13.

Multiply	x	-4
5x	$5x(x) = 5x^2$	$5x(-4) = -20x$
-1	-x	$-1(-4) = +4$

$(5x - 1)(x - 4) = 5x^2 - 20x - x + 4$

$= 5x^2 - 21x + 4$

15. a. $(x - 2)(x - 2) = x^2 - 2x - 2x + 4$

$= x^2 - 4x + 4$

b. $(x + 2)(x - 2) = x^2 - 2x + 2x - 4$

$= x^2 - 4$

17. a. $(a + 5)(a + 5) = a^2 + 5x + 5a + 25$

$= a^2 + 10a + 25$

b. $(b + 5)(b - 5) = b^2 - 5b + 5b - 25$

$= b^2 - 25$

19. a. $(a + b)(a - b) = a^2 - ab + ab - b^2$

$= a^2 - b^2$

b. $(a - b)(a - b) = a^2 - ab - ab + b^2$

$= a^2 - 2ab + b^2$

21. a. $(x + 1)(x + 7) = x^2 + 7x + x + 7$

$= x^2 + 8x + 7$

b. $(x + 1)(x - 7) = x^2 - 7x + x - 7$

$= x^2 - 6x - 7$

Section 6.2 (con't)

23. a. $(b + 7)(b + 7) = b^2 + 7b + 7b + 49$

$= b^2 + 14b + 49$

b. $(a + 7)(a - 7) = a^2 + 7a - 7a - 49$

$= a^2 - 49$

25. a. $(x + y)(x + y) = x^2 + xy + xy + y^2$

$= x^2 + 2xy + y^2$

b. $(x - y)(x - y) = x^2 - xy - xy + y^2$

$= x^2 - 2xy + y^2$

27. Perfect square trinomials are $(a + b)^2 =$ $a^2 + 2ab + b^2$. Exercises 15a, 16a, 17a, 18b, 19b, 20b, 23a, 24b, 25a and 25b are all perfect square trinomials.

29. Perfect square trinomial,

$(2x + 3)^2 = (2x)^2 + 2(2x)(3) + (3)^2$

$= 4x^2 + 12x + 9$

31. Difference of squares,

$(2x - 3)(2x + 3) = (2x)^2 - (3)^2$

$= 4x^2 - 9$

33. Neither,

$(2x - 3)(3 - 2x) = 6x - 4x^2 - 9 + 6x$

$= -4x^2 + 12x - 9$

35. Perfect square trinomial,

$(3x - {'}2)^2 = (3x)^2 + 2(3x)(-2) + (-2)^2$

$= 9x^2 - 12x + 4$

37. Difference of squares,

$(3x + 2)(3x - 2) = (3x)^2 - (2)^2$

$= 9x^2 - 4$

39. $(x + 5)^2 = x^2 + 2(5x) + 25$

$= x^2 + 10x + 25$

41. $(a - 6)^2 = a^2 + 2(-6a) + (-6)^2$

$= a^2 - 12a + 36$

43. $2(x + 3)^2 = 2[x^2 + 2(3x) + 9]$

$= 2(x^2 + 6x + 9) = 2x^2 + 12x + 18$

45. $3(x - 5)^2 = 3[x^2 + 2(-5x) + (-5)^2]$

$= 3(x^2 - 10x + 25) = 3x^2 - 30x + 75$

47. $(x - a)^2 = (x - a)(x - a)$, exponent should not be distributed.

49. $(x - 1)(x + 12) = x^2 + (-1 + 12)x - 12$

$= x^2 + 11x - 12$

51. $(x - 3)(x + 4) = x^2 + (-3 + 4)x - 12$

$= x^2 + 1x - 12$

53. $(x + 2)(x - 6) = x^2 + (2 - 6)x - 12$

$= x^2 - 4x - 12$

55. a. $(x + 1)(x + 8) = x^2 + (1 + 8)x + 8$

$= x^2 + 9x + 8$

b. $(x + 1)(x - 8) = x^2 + (1 - 8)x - 8$

$= x^2 - 7x - 8$

Section 6.2 (con't)

57. a. $(x + 2)(x + 4) = x^2 + (2 + 4)x + 8$

$= x^2 + 6x + 8$

b. $(x + 2)(x - 4) = x^2 + (2 - 4)x - 8$

$= x^2 - 2x - 8$

59. a. $(2x - 3)(3x - 2) = 6x^2 - 4x - 9x + 6$

$= 6x^2 - 13x + 6$

b. $(2x + 3)(3x - 2) = 6x^2 - 4x + 9x - 6$

$= 6x^2 + 5x - 6$

61. a. $(6x + 1)(x + 6) = 6x^2 + 36x + x + 6$

$= 6x^2 + 37x + 6$

b. $(6x - 1)(x + 6) = 6x^2 + 36x - x - 6$

$= 6x^2 + 35x - 6$

63. a. $(2x + 5)(x + 1) = 2x^2 + 2x + 5x + 5$

$= 2x^2 + 7x + 5$

b. $(2x + 5)(x - 1) = 2x^2 - 2x + 5x - 5$

$= 2x^2 + 3x - 5$

65. a. $(2x - 1)(x + 5) = 2x^2 + 10x - x - 5$

$= 2x^2 + 9x - 5$

b. $(2x + 1)(x + 5) = 2x^2 + 10x + x + 5$

$= 2x^2 + 11x + 5$

67. The constant terms must multiply to +12. The six sets are: $(x + 1)(x + 12)$;

$(x - 1)(x - 12)$; $(x + 2)(x + 6)$;

$(x - 2)(x - 6)$; $(x + 3)(x + 4)$;

$(x - 3)(x - 4)$. ✓

69. The constant terms must multiply to +20. The six sets are: $(x + 1)(x + 20)$;

$(x - 1)(x - 20)$; $(x + 2)(x + 10)$;

$(x - 2)(x - 10)$; $(x + 4)(x + 5)$;

$(x - 4)(x - 5)$

71. The value of n will be the sum of 2 numbers that multiply to +24. Possible values for n are: $1 + 24 = 25$; $2 + 12 = 14$; $3 + 8 = 11$; $4 + 6 = 10$. Sum

73. The coefficient -2 is obtained by multiplying 3 times x and -5 times x and then adding the like terms. $3x - 5x = -2x$

Mid-Chapter 6 Test

1. a. $7b - 8c + 3a + 4b - 5c - 6a$

$= (3 - 6)a + (7 + 4)b + (-8 - 5)c$

$= -3a + 11b - 13c$; trinomial

b. $x^2 - 4x + x^2 - 6 = 2x^2 - 4x - 6$,

trinomial

c. $4xy^2 + x^3y^2 - 3x^2y + x^3y^2 - 2xy^2$

$= 2x^3y^2 - 3x^2y + (4 - 2)xy^2$

$= 2x^3y^2 - 3x^2y + 2xy^2$, trinomial

d. $(5a - 3b - 2c) - (3a + 4b - 6c)$

$= 5a - 3b - 2c - 3a - 4b + 6c$

$= (5 - 3)a + (-3 - 4)b + (-2 + 6)c$

$= 2a - 7b + 4c$, trinomial

e. $x(x^2 + 5x + 25) - 5(x^2 + 5x + 25)$

$= x^3 + 5x^2 + 25x - 5x^2 - 25x - 125$

$= x^3 - 125$, binomial

f. $9 - 3x(x + 3) = 9 - 3x^2 - 9x$

$= -3x^2 - 9x + 9$, trinomial

g. $8 - 4(4 - x) = 8 - 16 + 4x$

$= 4x - 8$, binomial

2. a. gcf = 2, $6x^2 - 2x + 8 = 2(3x^2 - x + 4)$

b. gcf = a,

$2abc - 3ac + 4ab = a(2bc - 3c + 4b)$

3. a. Length = $2a + b$, Width = $a + 2b$

$P = 2(2a + b) + 2(a + 2b)$,

$P = 4a + 2b + 2a + 4b$, $P = 6a + 6b$

$A = (2a + b)(a + 2b)$,

$A = 2a^2 + 4ab + ab + 2b^2$, $A = 2a^2 + 5ab$,

$A = 2b^2$

b. Length = $x + 2$, Width = $x + 1$

$P = 2(x + 2) + 2(x + 1)$,

$P = 2x + 4 + 2x + 2$, $P = 4x + 6$

$A = (x + 2)(x + 1)$, $A = x^2 + 3x + 2$

4. Terms are added (or subtracted) to each
other $(a + b)$, factors are multiplied (ab).

5. $(a + 2b)(3a + 2b)$

6.

Multiply	2x	-5
3x	$3x(2x) = 6x^2$	$3x(-5) = -15x$
+2	$+2(2x) = +4x$	$+2(-5) = -10$

$(3x + 2)(2x - 5) = 6x^2 - 15x + 4x - 10$

$= 6x^2 - 11x - 10$

7. $(-15x)(+4x) = -60x^2$

8.

Multiply	$3x^2$	-2x	+1
x	$3x^3$	$-2x^2$	$+x$
-2	$-6x^2$	$+4x$	-2

$(x - 2)(3x^2 - 2x + 1) = 3x^3 - 8x^2 + 5x - 2$

Mid-Chapter 6 Test (con't)

9. Like terms are $-6x^2$ and $-2x^2$; $4x$ and x

10. $(x + 3)(x + 5) = x^2 + 5x + 3x + 15$

    $= x^2 + 8x + 15$

11. $(x - 4)(3x + 5) = 3x^2 + 5x - 12x - 20$

    $= 3x^2 - 7x - 20$

12. $(2x + 7)(3x - 1) = 6x^2 - 2x + 21x - 7$

    $= 6x^2 + 19x - 7$

13. $(x - 3)(x + 3) = x^2 - 9$

14. $(2x - 5)^2 = (2x)^2 + 2(-10x) + (-5)^2$

    $= 4x^2 - 20x + 25$

15. a. Neither

    b. Neither

    c. Difference of squares

    d. Perfect square trinomial

16. $(x \pm 1)(x \pm 10) = x^2 \pm 11x + 10,$

    $(x \mp 1)(x \pm 10) = x^2 \pm 9x - 10,$

    $(x \pm 2)(x \pm 5) = x^2 \pm 7x + 10,$

    $(x \mp 2)(x \pm 5) = x^2 \pm 3x - 10$

Section 6.3

1. $(2x + 1)(x + 2) = 2x^2 + 5x + 2$

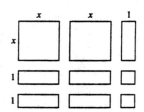

3. $(2x + 1)(2x + 1) = 4x^2 + 4x + 1$

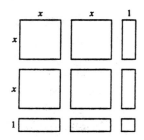

5.

Factor	x	+5
x	x^2	+5x
+4	+4x	+20

$(x + 5)(x + 4) = x^2 + 9x + 20$

7.

Factor	x	+2
x	x^2	+2x
-10	-10x	-20

$(x + 2)(x - 10) = x^2 - 8x - 20$

9.

Factor	6x	+1
x	$6x^2$	+x
-3	-18x	-3

$(6x + 1)(x - 3) = 6x^2 - 17x - 3$

11. The coefficient -2 is obtained by multiplying 3 times x and -5 times x and then adding the like terms. $3x - 5x = -2x$

13. $x^2 + 8x + 12 = x^2 + 2x + 6x + (2)(6)$

$= (x + 2)(x + 6)$

15. $x^2 - 13x + 12 = x^2 - x - 12x + (-1)(-12)$

$= (x - 1)(x - 12)$

17. $x^2 + x + 12$ can not be factored

19. $x^2 + 7x + 12 = x^2 + 3x + 4x + (3)(4)$

$= (x + 3)(x + 4)$

21. $x^2 + x - 12 = x^2 + 4x - 3x + (4)(-3)$

$= (x + 4)(x - 3)$

23. $x^2 - 11x - 12 = x^2 - 12x + x + (-12)(1)$

$= (x - 12)(x + 1)$

25. $x^2 + 6x + 9$ is a perfect square trinomial, factors to $(x + 3)^2$

27. $x^2 + 11x + 30 = x^2 + 5x + 6x + (5)(6)$

$= (x + 5)(x + 6)$

29. $x^2 + 13x - 30 = x^2 + 15x - 2x + (15)(-2)$

$= (x + 15)(x - 2)$

Section 6.3 (con't)

31. $x^2 - 6x - 16 = x^2 + 2x - 8x + (2)(-8)$

$= (x + 2)(x - 8)$

33. $x^2 + 15x - 16 = x^2 + 16x - x + (16)(-1)$

$= (x + 16)(x - 1)$

35. $x^2 - 25$ is a difference of squares, factors

to $(x + 5)(x - 5)$

37. $x^2 - 4 = (x + 2)(x - 2)$, difference of

squares

39. $4x^2 - 16 = 4(x^2 - 4) = 4(x + 2)(x - 2)$,

difference of squares

41. $x^2 + 12x + 36 = (x + 6)^2$, perfect square

trinomial

43. $4x^2 + 8x + 4 = 4(x^2 + 2x + 1)$

$= 4(x + 1)^2$, perfect square trinomial

45. $x^2 + 4$, can not be factored

47. $x^2 - 6x + 9 = (x - 3)^2$, perfect square

trinomial

49. Diagonal product $= 24x^2$, sum $= 11x$;

middle terms are 3x and 8x

Factor	2x	+3
x	$2x^2$	3x
+4	8x	+12

$2x^2 + 11x + 12 = (x + 4)(2x + 3)$

51. Diagonal product $= -18x^2$, sum $= -3x$,

middle terms are 3x and -6x

Factor	2x	+3
x	$2x^2$	3x
-3	-6x	-9

$2x^2 - 3x - 9 = (x - 3)(2x + 3)$

53. Diagonal product $= -6n^2$, sum $= +n$,

middle terms are 3n and -2n

Factor	2n	+3
n	$2n^2$	3n
-1	-2n	-3

$2n^2 + n - 3 = (n - 1)(2n + 3)$

55. Diagonal product $= -6x^2$, sum $= +5x$,

middle terms are 6x and -x

Factor	3x	-1
x	$3x^2$	-x
+2	6x	-2

$3x^2 + 5x - 2 = (x + 2)(3x - 1)$

57. Diagonal product $= -12a^2$, sum $= -11a$,

middle terms are +a and -12a

Factor	3a	+1
a	$3a^2$	+a
-4	-12a	-4

$3a^2 - 11a - 4 = (a - 4)(3a + 1)$

59. $3a^2 + 5a - 4$ can not be factored

Diagonal product $= -12a^2$, sum $= +5a$

Section 6.3 (con't)

61. $3x^2 + 10x - 5$ can not be factored

Diagonal product = $-15x^2$, sum = $+10x$

63. $9x^2 - 49 = (3x + 7)(3x - 7)$, difference of squares

65. $16x^2 - 9 = (4x + 3)(4x - 3)$, difference of squares

67. Diagonal product = $-12x^2$, sum = $+x$,

middle terms are $4x$ and $-3x$

Factor	3x	+2
2x	$6x^2$	+4x
-1	-3x	-2

$6x^2 + x - 2 = (2x - 1)(3x + 2)$

69. Diagonal product = $-36x^2$, sum = $+5x$,

middle terms are $9x$ and $-4x$

Factor	3x	-2
2x	$6x^2$	-4x
+3	9x	-6

$6x^2 + 5x - 6 = (2x + 3)(3x - 2)$

71. Diagonal product = $-10n^2$, sum = $+9n$,

middle terms are $10n$ and $-n$

Factor	n	+5
2n	$2n^2$	10n
-1	-n	-5

$2n^2 + 9n - 5 = (2n - 1)(n + 5)$

73. $25x^2 - 36 = (5x + 6)(5x - 6)$, difference of squares

75. Remove common factor of 3 first,

Diagonal product = $3x^2$, sum = $+4x$,

middle terms are x and $3x$

Factor	x	+1
x	x^2	+x
+3	3x	+3

$3x^2 + 12x + 9 = 3(x^2 + 4x + 3)$

$= 3(x + 3)(x + 1)$

77. $3x^2 - 27 = 3(x^2 - 9) = 3(x + 3)(x - 3)$, difference of squares

79. $5x^2 - 10x + 5 = 5(x^2 - 2x + 1) = 5(x - 1)^2$, perfect square trinomial

81. $18x^2 - 50 = 2(9x^2 - 25) = 2(3x + 5)(3x - 5)$ difference of squares

83. $3x^2 - 30x + 75 = 3(x^2 - 10x + 25)$

$= 3(x - 5)^2$, perfect square trinomial

85. $3x^2 + 6x + 12 = 3(x^2 + 2x + 4)$

Section 6.4

1.

Input: x	Output:3^x
2	$3^2 = 9$
1	$3^1 = 3$
0	$3^0 = 1$
-1	$3^{-1} = \frac{1}{3}$
-2	$3^{-2} = \frac{1}{3^2} = \frac{1}{9}$
-3	$3^{-3} = \frac{1}{3^3} = \frac{1}{27}$

3.

Input: x	Output:4^x
2	$4^2 = 16$
1	$4^1 = 4$
0	$4^0 = 1$
-1	$4^{-1} = \frac{1}{4}$
-2	$4^{-2} = \frac{1}{4^2} = \frac{1}{16}$
-3	$4^{-3} = \frac{1}{4^3} = \frac{1}{64}$

5. a. $2^{-2} = \dfrac{1}{2^2} = \dfrac{1}{4}$

　　b. $2^{-1} = \dfrac{1}{2}$

　　c. $2^0 = 1$

7. a. $\left(\frac{1}{4}\right)^0 = 1$

　　b. $\left(\frac{1}{4}\right)^{-2} = 4^2 = 16$

　　c. $\left(\frac{1}{4}\right)^{-1} = 4^1 = 4$

9. a. $(0.5)^{-1} = \left(\frac{1}{2}\right)^{-1} = 2^1 = 2$

　　b. $(0.5)^0 = 1$

　　c. $(0.5)^{-2} = \left(\frac{1}{2}\right)^{-2} = 2^2 = 4$

11. An exponent of -1 gives the reciprocal of a number.

13. An exponent of zero gives 1 when the base is not equal to zero.

15. Answers will vary, (0.5) and $(0.5)^{-1}$ is one example.

17. a. $5^3 \cdot 5^{-7} = 5^{3+(-7)} = 5^{-4}$

　　b. $6^5 \cdot 6^{-2} = 6^{5+(-2)} = 6^3$

　　c. $10^3 \cdot 10^{-6} = 10^{3+(-6)} = 10^{-3}$

　　d. $10^2 \cdot 10^{-7} = 10^{2+(-7)} = 10^{-5}$

19. a. $10^{-15} \cdot 10^{-15} = 10^{-15+(-15)} = 10^{-30}$

　　b. $10^{-28} \cdot 10^{19} = 10^{-28+19} = 10^{-9}$

　　c. $2^{24} \cdot 2^{-16} = 2^{24+(-16)} = 2^8$

　　d. $2^{13=} \cdot 2^{-5} = 2^{13+(-5)} = 2^8$

Section 6.4 (con't)

21. a. $\dfrac{3^5}{3^{-2}} = 3^{5-(-2)} = 3^7$

b. $\dfrac{10^{-5}}{10^{-2}} = 10^{-5-(-2)} = 10^{-3}$

c. $\dfrac{10^{-4}}{10^{12}} = 10^{-4-12} = 10^{-16}$

d. $\dfrac{1}{10^2} = 10^{-2}$

23. a. $(2^3)^4 = 2^{3 \cdot 4} = 2^{12}$

b. $(2^3)^{-4} = 2^{3(-4)} = 2^{-12}$

c. $(10^5)^2 = 10^{5 \cdot 2} = 10^{10}$

d. $(10^{-6})^3 = 10^{(-6) \cdot 3} = 10^{-18}$

25. a $a^5 \cdot a^{-12} = a^{5+(-12)} = a^{-7}$

b. $x^{-5} \cdot x^{13} = x^{-5+13} = x^8$

c. $n^{-6} \cdot n^2 = n^{-6+2} = n^{-4}$

27. a. $\dfrac{1}{x^2} = x^{-2}$

b. $\dfrac{1}{a^{-1}} = a$

c. $\dfrac{b}{b^{-2}} = b^{1-(-2)} = b^3$

29. a. $\dfrac{a^3}{a^{-6}} = a^{3-(-6)} = a^9$

b. $\dfrac{a^{-6}}{a^2} = a^{-6-2} = a^{-8}$

c. $\dfrac{x^4}{x^{-2}} = x^{4-(-2)} = x^6$

31. a. $(x^2)^{-4} = x^{2 \cdot (-4)} = x^{-8}$

b. $(x^{-4})^{-3} = x^{-4 \cdot (-3)} = x^{12}$

c. $(b^{-2})^3 = b^{-2 \cdot 3} = b^{-6}$

33. a. $x^{-1} = \dfrac{1}{x}$

b. $\left(\dfrac{x}{y}\right)^{-1} = \dfrac{y}{x}$

c. $\left(\dfrac{y}{x}\right)^{-1} = \dfrac{x}{y}$

d. $\left(\dfrac{a}{b}\right)^0 = 1$

e. $\left(\dfrac{a}{c}\right)^0 = 1$

f. $\left(\dfrac{a}{bc}\right)^{-1} = \dfrac{bc}{a}$

Section 6.4 (con't)

35. a $y^{-3} = \dfrac{1}{y^3}$

b. $\left(\dfrac{y}{x}\right)^{-2} = \left(\dfrac{x}{y}\right)^2 = \dfrac{x^2}{y^2}$

c. $\dfrac{1}{b^{-3}} = b^3$

d. $\left(\dfrac{a}{b}\right)^{-3} = \left(\dfrac{b}{a}\right)^3 = \dfrac{b^3}{a^3}$

e. $\left(\dfrac{4a^2}{c}\right)^{-2} = \left(\dfrac{c}{4a^2}\right)^2 = \dfrac{c^2}{4^2(a^2)^2} = \dfrac{c^2}{16a^4}$

f. $\left(\dfrac{a}{b^2}\right)^{-3} = \left(\dfrac{b^2}{a}\right)^3 = \dfrac{b^{2\cdot3}}{a^3} = \dfrac{b^6}{a^3}$

37. a. $\dfrac{xy^{-2}}{x^2y^3} = x^{1-2}y^{-2-3} = x^{-1}y^{-5} = \dfrac{1}{xy^5}$

b. $\dfrac{x^{-1}y}{x^{-1}y^{-2}} = x^{-1-(-1)}y^{1-(-2)} = x^0y^3 = y^3$

c. $\dfrac{a^{-2}b^2}{a^3b^{-1}} = a^{-2-3}b^{2-(-1)} = a^{-5}b^3 = \dfrac{b^3}{a^5}$

39. a. $1(2x)^2(-3)^0 + 2(2x)^1(-3)^1 + 1(2x)^0(-3)^2$

$= 2^2x^2\cdot1 - 12x + 1\cdot(-3)^2 = 4x^2 - 12x + 9$

b. $1(2x)^3(3)^0 + 3(2x)^2(3)^1 + 3(2x)^1(3)^2 + 1(2x)^0(3)^3$

$= 2^3x^3\cdot1 + 9\cdot2^2x^2 + 6x\cdot3^2 + 1\cdot3^3$

$= 8x^3 + 36x^2 + 54x + 27$

c. $1(3x)^3(-2)^0 + 3(3x)^2(-2)^1 + 3(3x)^1(-2)^2 + 1(3x)^0(-2)^3$

$= 3^3x^3\cdot1 - 6\cdot3^2x^2 + 9x\cdot(-2)^2 + 1\cdot(-2)^3$

$= 27x^3 - 54x^2 + 36x - 8$

41. a. $(x^{-1})^{-1} = x^{(-1)\cdot(-1)} = x$

b. $\dfrac{1}{x^{-1}} = x^1 = x$

c. $\dfrac{1}{\frac{1}{x}} = 1\cdot\dfrac{x}{1} = x$

All the expressions equal x.

43.

x	10^x as fraction	10^x as decimal
0	1	1
-1	$\frac{1}{10}$	0.1
-2	$\frac{1}{100}$	0.01
-3	$\frac{1}{1000}$	0.001
-4	$\frac{1}{10000}$	0.0001
-5	$\frac{1}{100000}$	0.00001

Section 6.4 (con't)

45. a. $5280^3 = 1.47 \times 10^{11}$

    b. $0.00008^3 = 5.12 \times 10^{-13}$

47. a. $2.34\ -02 = 2.34 \times 10^{-2} = 0.0234$

    b. $3.14\ 03 = 3.14 \times 10^3 = 3140$

    c. $6.28\ 07 = 6.28 \times 10^7 = 62,800,000$

49. $1,391,400 \approx 1.39 \times 10^6$ km

51. 2756.4 million $= 2756.4 \times 10^6$

    $\approx 2.76 \times 10^9$ miles

53. $1800 = 1.80 \times 10^3$ g

55. a. $34 \times 10^3 = 3.4 \times 10^4$

    b. $560 \times 10^{-2} = 5.60$

57. a. $0.432 \times 10^4 = 4.32 \times 10^3$

    b. $0.567 \times 10^{-5} = 5.67 \times 10^{-6}$

59. $1,990,000,000,000,000,000,000,000,000,000$ kg

61. $-0.0000000000000000001602$ Coulombs

63. $200,000,000$ years

65. $0.00000000000000000000000000016750$ kg

67. a. $2 \times 10^5 = 200,000;\ 3 \times 10^4 = 30,000$

    $(2 \times 10^5) > (3 \times 10^4)$

    b. $4 \times 10^3 = 4,000;$

    $(4 \times 10^3) < (3 \times 10^4)$

69. a. $2 \times 10^{-5} = 0.00002,\ 3 \times 10^{-4} = 0.0003$

    $(2 \times 10^{-5}) < (3 \times 10^{-4})$

    b. $4 \times 10^{-3} = 0.004$

    $(4 \times 10^{-3}) > (3 \times 10^{-4})$

71. a. $3.2 > 2.8$ and $10^{-14} = 10^{-14}$ so:

    $(3.2 \times 10^{-14}) > (2.8 \times 10^{-14})$

    b. $4.3 < 5.3$ and $10^{-13} = 10^{-13}$ so:

    $(4.3 \times 10^{-13}) < (5.3 \times 10^{-13})$

73. The electron has a smaller mass.

    $9.101 \times 10^{-31} < 1.6726 \times 10^{-27}$

75. From the text we know that light travels 5.87×10^{12} miles in one light year.

    $(5.87 \times 10^{12})(27,000) = 158490 \times 10^{12}$

    $\approx 1.58 \times 10^{17}$

77. $18,700$ tons $\cdot \dfrac{2000\ \text{lb}}{1\ \text{ton}} \cdot \dfrac{1\ \text{gal}}{8.3\ \text{lb}} \cdot \dfrac{4\ \text{qt}}{1\ \text{gal}}$

    $\approx 1.8 \times 10^7$ qts of water

79. 1.36×10^9 lb $\cdot \dfrac{16\ \text{oz}}{1\ \text{lb}} \cdot \dfrac{160\ \text{cal}}{1\ \text{oz}}$

    $\approx 3.48 \times 10^{12}$ cal.

    1 billion $= 10^9$, $(3.48 \times 10^{12}) \div 5.6 \times 10^9$

    $= (3.48 \div 5.6) \times 10^{12-9} \approx 0.621 \times 10^3$

    ≈ 621 calories per person

Section 6.4 (con't)

81. 5 ft 9 inches = 69 inches

69 - 20 = 49 inches growth

$$\frac{49\,in}{18\,yr}\cdot\frac{1\,yr}{365\,days}\cdot\frac{1\,day}{24\,hr}\cdot\frac{1\,ft}{12\,in}\cdot\frac{1\,mile}{5280\,ft}$$

$\approx 4.90 \times 10^{-9}$ mph

83. a. $(2 \times 10^{15})(3 \times 10^{12}) = (2)(3) \times 10^{15+12}$

$= 6 \times 10^{27}$

b. $(4 \times 10^{13})(3 \times 10^{18}) = (4)(3) \times 10^{13+18}$

$= 12 \times 10^{31} = 1.2 \times 10^{32}$

85. a. $(5 \times 10^{-16})(6 \times 10^{-11})$

$= (5)(6) \times 10^{-16+(-11)} = 30 \times 10^{-27}$

$= 3 \times 10^{-26}$

b. $(8 \times 10^{-12})(6 \times 10^{-13})$

$= (8)(6) \times 10^{-12+(-13)} = 48 \times 10^{-25}$

$= 4.8 \times 10^{-24}$

87. a. $(2.8 \times 10^{-14}) \div (7 \times 10^{-18})$

$= (2.8 \div 7) \times 10^{-14-(-18)} = 0.4 \times 10^4$

$= 4 \times 10^3$

b. $(9 \times 10^{-25}) \div (4.5 \times 10^{-12})$

$= (9 \div 4.5) \times 10^{-25-(-12)} = 2 \times 10^{-13}$

89. $\dfrac{4.8 \times 10^{-7}}{(1.6 \times 10^8)(3.0 \times 10^{-18})}$

$= \dfrac{4.8 \times 10^{-7}}{4.8 \times 10^{-10}} = 1 \times 10^{-7-(-10)} = 1 \times 10^3$

Chapter 6 Review

1. a. $3a^2 - 5ab + 4a^2 - 3b^2 + 2ab - 7b^2$

$= (3 + 4)a^2 + (-5 + 2)ab + (-3 - 7)b^2$

$= 7a^2 - 3ab - 10b^2$, trinomial

b. $3 + 4x^2 + 5x - 2(x - 1)$

$= 4x^2 + 5x - 2x + 2 + 3 = 4x^2 + 3x + 5$,
trinomial

c. $x(x^2 + 4x + 16) - 4(x^2 + 4x + 16)$

$= x^3 + 4x^2 + 16x - 4x^2 - 16x - 64$

$= x^3 - 64$, binomial

d. $11x - 4x(1 - x) = 11x - 4x + 4x^2$

$= 4x^2 + 7x$, binomial

e. $9 - 4(x - 3) = 9 - 4x + 12$

$= -4x + 21$, binomial

3. a. $P = 2(2x) + 2(3x + 1), P = 4x + 6x + 2$,

$P = 10x + 2$

$A = 2x(3x + 1), A = 6x^2 + 2x$

b. $P = 2(x + 2) + 2(3x), P = 2x + 4 + 6x$,

$P = 8x + 4$

$A = 3x(x + 2), A = 3x^2 + 6x$

c. $P = (3x - 4) + 2x + (2x + 1)$

$P = 3x + 2x + 2x - 4 + 1, P = 7x - 3$

$A = \frac{1}{2}(2x + 1)(2x), A = (2x + 1)x$,

$A = 2x^2 + x$

5. $(3x + 1)(x + 1) = 3x^2 + 4x + 1$

7. $(x + 4)(x + 3) = x^2 + 3x + 4x + 12$

$= x^2 + 7x + 12$

9. $(2x - 5)(2x - 5) = (2x)^2 + 2(2x)(-5) + (-5)^2$

$= 4x^2 - 20x + 25$

11. $(3x + 5)(3x - 2) = 9x^2 - 6x + 15x - 10$

$= 9x^2 + 9x - 10$

13. $(2x - 3)(3x + 2) = 6x^2 + 4x - 9x - 6$

$= 6x^2 - 5x - 6$

15. $(a - b)(a - b) = a^2 - 2ab + b^2$

17. Exercises 9 and 15 are perfect square
trinomials.

19. a.

Factor	x	+7
x	x^2	+7x
-2	-2x	-14

$(x + 7)(x - 2) = x^2 + 5x - 14$

b.

Factor	3x	+5
2x	$6x^2$	+10x
+3	+9x	+15

$(3x + 5)(2x + 3) = 6x^2 + 19x + 15$

Chapter 6 Review (con't)

21. a. Diagonal product = $14x^2$, sum = $-9x$

Factor	x	-2
x	x^2	-2x
-7	-7x	+14

$(x - 2)(x - 7) = x^2 - 9x + 14$

b. Diagonal product = $-42x^2$, sum = $11x$,

Factor	x	+7
2x	$2x^2$	+14x
-3	-3x	-21

$(x + 7)(2x - 3) = 2x^2 + 11x - 21$

c. Diagonal product = $-12x^2$, sum = $-x$,

Factor	3x	-1
4x	$12x^2$	-4x
+1	+3x	-1

$(3x - 1)(4x + 1) = 12x^2 - x - 1$

d. Diagonal product = $-180x^2$,

sum = $24x$

Factor	2x	+3
10x	$20x^2$	+30x
-3	-6x	-9

$(2x + 3)(10x - 3) = 20x^2 + 24x - 9$

23. $x^2 - 3x + 2 = x^2 - x - 2x + 2$

$= (x - 1)(x - 2)$, neither

25. $9x^2 - 16 = (3x + 4)(3x - 4)$, ds

27. $2x^2 + x - 6 = 2x^2 + 4x - 3x - 6$

$= (2x - 3)(x + 2)$, neither

29. $9x^2 + 3x - 2 = 9x^2 + 6x - 3x - 2$

$= (3x + 2)(3x - 1)$, neither

31. $x^2 + 6x + 8 = x^2 + 4x + 2x + 8$

$= (x + 4)(x + 2)$, neither

33. $x^2 - 11x + 10 = x^2 - x - 10x + 10$

$= (x - 1)(x - 10)$, neither

35. $25 - 9x^2 = (5 + 3x)(5 - 3x)$, ds

37. $y^2 + 8y + 12 = y^2 + 6y + 2y + 12$

$= (y + 6)(y + 2)$, neither

39. $2x^2 - 3x - 35 = 2x^2 - 10x + 7x - 35$

$= (2x + 7)(x - 5)$, neither

41. $4x^2 - 8x + 4 = 4(x^2 - 2x + 1)$

$= 4(x - 1)^2$, pst

43. $x^3 + 4x^2 + 4x = x(x^2 + 4x + 4)$

$= x(x + 2)^2$, contains pst

45. $3x^2 - 27 = 3(x^2 - 9)$

$= 3(x + 3)(x - 3)$, contains ds

47. $x^3 - 7x^2 + 10x = x(x^2 - 7x + 10)$

$= x(x^2 - 2x - 5x + 10) = x(x - 2)(x - 5)$,

neither

49. a. $3^{-1} = \frac{1}{3}$

b. $3^0 = 1$

c. $3^{-2} = \frac{1}{3^2} = \frac{1}{9}$

Chapter 6 Review (con't)

51. a. $\left(\frac{2}{3}\right)^0 = 1$

    b. $\left(\frac{2}{3}\right)^{-1} = \frac{3}{2}$

    c. $\left(\frac{2}{3}\right)^{-2} = \left(\frac{3}{2}\right)^2 = \frac{9}{4}$

53. a. $x^7 x^{-2} = x^{7+(-2)} = x^5$

    b. $x^3 x^{-3} = x^{3+(-3)} = x^0 = 1$

    c. $b^{-5} b^{-5} = b^{-5+(-5)} = \dfrac{1}{b^{10}}$

55. a. $\dfrac{n^4}{n^{-5}} = n^{4-(-5)} = n^9$

    b. $\dfrac{n^{-4}}{n^{-5}} = n^{-4-(-5)} = n$

57. a. $(b^2)^{-4} = b^{2\cdot(-4)} = b^{-8} = \dfrac{1}{b^8}$

    b. $\left(x^{-2}\right)^{-3} = x^{(-2)(-3)} = x^6$

    c. $\left(b^0\right)^{-2} = b^{0\cdot(-2)} = b^0 = 1$

59. a. $\left(\dfrac{a^2}{b^2}\right)^{-1} = \dfrac{b^2}{a^2}$

    b. $\left(\dfrac{2a}{b^2 c}\right)^{-3} = \left(\dfrac{b^2 c}{2a}\right)^3 = \dfrac{b^6 c^3}{8a^3}$

61. a. $\dfrac{x^{-3} y^{-3}}{x^5 y^{-2}} = x^{-3-5} y^{-3-(-2)} = x^{-8} y^{-1} = \dfrac{1}{x^8 y}$

    b. $\dfrac{a^3 b^{-6}}{a^6 b^{-4}} = a^{3-6} b^{-6-(-4)} = a^{-3} b^{-2}$

    $= \dfrac{1}{a^3 b^2}$

63. a. $\dfrac{x^3 x^0}{x^0} = x^{3+0-0} = x^3$

    b. $\dfrac{a^{-2} a^2}{a^0} = a^{-2+2-0} = a^0 = 1$

65. Potassium, 1,400,000,000 yrs

    Calcium, 120,000 yrs

    Radon, 0.0000001243 yr

    Polonium, 0.00000030 sec

67. $3.0 \times 10^7 \text{ sec} \cdot \dfrac{1\,\text{min}}{60\,\text{sec}} \cdot \dfrac{1\,\text{hr}}{60\,\text{min}} \cdot \dfrac{1\,\text{day}}{24\,\text{hr}} \cdot \dfrac{1\,\text{yr}}{365\,\text{day}}$

    $\approx 9.5 \times 10^{-15}$ yr

69. a. $(6.0 \times 10^8)(7.0 \times 10^{-2})$

    $= (6.0)(7.0) \times 10^{8+(-2)} = 42 \times 10^6$

    $= 4.2 \times 10^7$

    b. $(8.0 \times 10^7)(4.0 \times 10^{-3})$

    $= (8.0)(4.0) \times 10^{7+(-3)} = 32 \times 10^4$

    $= 3.2 \times 10^5$

Chapter 6 Review (con't)

71. a. $\dfrac{7.2 \times 10^9}{8.0 \times 10^{-8}} = \dfrac{7.2}{8.0} \times 10^{9-(-8)}$

   $= 0.9 \times 10^{17} = 9 \times 10^{16}$

b. $\dfrac{5.4 \times 10^9}{9.0 \times 10^{-6}} = \dfrac{5.4}{9.0} \times 10^{9-(-6)}$

   $= 0.6 \times 10^{15} = 6 \times 10^{14}$

73. a. $234 \times 10^5 = 2.34 \times 10^7$

b. $0.0436 \times 10^8 = 4.36 \times 10^6$

75. $\dfrac{2\,\text{in}}{15\,\text{min}} \cdot \dfrac{60\,\text{min}}{1\,\text{hr}} \cdot \dfrac{1\,\text{ft}}{12\,\text{in}} \cdot \dfrac{1\,\text{mile}}{5280\,\text{ft}}$

   $\approx 0.000126\ \text{mph} = 1.26 \times 10^{-4}\ \text{mph}$

Chapter 6 Test

1. 3 terms, $3x^2$, $4x$ and 1

2. Greatest common factor

3. Factoring

4. -1

5. $4^0 = 1$, $x = 0$

6. $\dfrac{1}{16} = \dfrac{1}{4^2} = 4^{-2}$, $x = -2$

7. 0.0003482

8. 4.5×10^{10}

9. $(a + 3b - 3c - d) + (3a - 5b + 8c - d)$

   $= a + 3a + 3b - 5b - 3c + 8c - d - d$

   $= 4a - 2b + 5c - 2d$, 4 term polynomial

10. $x(x^2 + 3x + 9) - 3(x^2 + 3x + 9)$

    $= x^3 + 3x^2 + 9x - 3x^2 - 9x - 27$

    $= x^3 - 27$, binomial

11. $-4x(x - 5) = -4x^2 + 20x$, binomial

12. $14xy + 6x^2y - 18y^2 = 6x^2y + 14xy - 18y^2$
    $= 2y(3x^2 + 7x - 9y)$

13.

Factor	$3x$	$+4$
$2x$	$6x^2$	$+8x$
-5	$-15x$	-20

$(3x + 4)(2x - 5) = 6x^2 - 7x - 20$

14. $P = 2(4x) + 2(2x + 5)$,

    $P = 8x + 4x + 10$, $P = 12x + 10$

    $A = 4x(2x + 5)$, $A = 8x^2 + 20x$

15. Let w = width;

    $6x^2 + 9x = w(2x + 3)$,

    $3x(2x + 3) = w(2x + 3)$,

    $w = 3x$

16. $(x - 4)(x + 7) = x^2 + 7x - 4x - 28$

    $= x^2 + 3x - 28$

17. $(x - 7)(x - 7) = x^2 + 2(-7x) + (-7)^2$

    $= x^2 - 14x + 49$

18. $(2x - 7)(2x + 7) = (2x)^2 - 7^2 = 4x^2 - 49$

19. $2(x - 4)(x - 4) = 2(x^2 - 8x -+ 16)$

    $= 2x^2 - 16x + 32$

20. $x^2 - 9x + 20 = x^2 - 4x - 5x + 20$

    $= (x - 4)(x - 5)$

21. $2x^2 - 3x - 2 = 2x^2 - 4x + x - 2$

    $= (2x + 1)(x - 2)$

22. $2x^2 - 8 = 2(x^2 - 4) = 2(x + 2)(x - 2)$

23. $x^2 - 8x + 16 = (x - 4)^2$

24. $9x^2 + 6x + 1 = (3x + 1)^2$

25. Exercises 18 and 22 are difference of squares.

Chapter 6 Test (con't)

26. a. $b^{-2}b^3 = b^{-2+3} = b$

   b. $\left(x^3\right)^{-2} = x^{3\cdot(-2)} = x^{-6} = \dfrac{1}{x^6}$

27. a. $\dfrac{b^3}{b^{-2}} = b^{3-(-2)} = b^5$

   b. $\dfrac{ab^0}{a^2b^{-1}} = a^{1-2}b^{0-(-1)} = a^{-1}b^1 = \dfrac{b}{a}$

28. a. $\left(\dfrac{a}{2b}\right)^0 = 1$

   b. $\left(\dfrac{9x^2}{25y^2}\right)^{-2} = \left(\dfrac{25y^2}{9x^2}\right)^2 = \dfrac{625y^4}{81x^4}$

29. $(2.5 \times 10^{-15})(4.0 \times 10^2)$

   $= (2.5)(4.0) \times 10^{-15+2} = 10 \times 10^{-13}$

   $= 1 \times 10^{-12} = 0.000000000001$

30. $\dfrac{1.25 \times 10^{-8}}{2.5 \times 10^{-4}} = \dfrac{1.25}{2.5} \times 10^{-8-(-4)} = 0.5 \times 10^{-4}$

   $= 5 \times 10^{-5}$

31. Factors $(x \pm 1)(x \pm 21)$; $(x \pm 3)(x \pm 7)$

   Trinomials $x^2 \pm 22x + 21$, $x^2 \pm 10x + 21$,

   $x^2 \pm 20x - 21$, $x^2 \pm 4x - 21$.

   1, 21 and 3, 7 are the only factors of 21.

32.

Multiply	a	+b
a	a^2	+ab
+b	+ab	b^2

There are two places in the table where a and b multiply each other resulting in the missing center term of 2ab.

33. 10 [EE] 3 is really $10 \times 10^3 = 10^{1+3} = 10^4$ $= 10,000$; 10 [y^x] 3 is $10^3 = 1000$.

Cumulative Review, Chapters 1 to 6

1.

Input: x	Output: y = 0.25x + 0.25
0	0.25(0) + 0.25 = 0.25
5	0.25(5) + 0.25 = 1.50
10	0.25(10) + 0.25 = 2.75
15	0.25(15) + 0.25 = 4.00
20	0.25(20) + 0.25 = 5.25
25	0.25(25) + 0.25 = 6.50
30	0.25(30) + 0.25 = 7.75

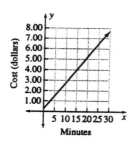

3. $5 - 2(4 - x) = 5 - 8 + 2x = 2x - 3$

5. $x^3 x^8 = x^{3+8} = x^{11}$

7. $R = \dfrac{(30)(20)}{30 + 20}$, $R = \dfrac{600}{50}$, $R = 12$

9. $3x - 13 = 13 - 2x$, $5x - 13 = 13$,

$5x = 26$, $x = 26 \div 5$, $x = 5.2$

11. $x + 1 \geq 9 - 3x$, $4x + 1 \geq 9$,

$4x \geq 8$, $x \geq 2$

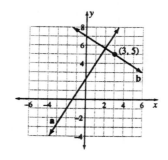

13. $3x + 5y = 15$, $5y = -3x + 15$,

$y = \dfrac{-3x + 15}{5}$, $y = -\frac{3}{5}x + 3$

15. $\dfrac{a + b + c}{3} = m$, $a + b + c = 3m$

$a = 3m - b - c$

17.

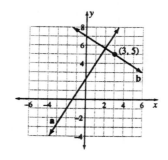

19. Slope from exercise 13 = $-\frac{3}{5}$, $b = 0$,

$y = -\frac{3}{5}x$

21. $\dfrac{1.19 - 0.92}{0.92} = \dfrac{0.27}{0.92} \approx 0.293$

29.3% increase

<u>**Cumulative Review (con't)**</u>

23. $\dfrac{x-1}{2} = \dfrac{x+3}{3}$, $3(x-1) = 2(x+3)$

    $3x - 3 = 2x + 6$, $x = 9$

25. $x(x^2 - 2x + 1) - (x^2 - 2x + 1)$

    $= x^3 - 2x^2 + x - x^2 + 2x - 1$

    $= x^3 - 3x^2 + 3x - 1$

27. $(2x + 1)(2x + 3) = 4x^2 + 6x + 2x + 3$

    $= 4x^2 + 8x + 3$

29. $4x^2 - 25 = (2x + 5)(2x - 5)$, difference of squares

Section 7.1

1. In $y = -2x$, $m = -2$, $b = 0$

    In $y = 1 - x$, $m = -1$, $b = 1$

    Intersection is at $(-1, 2)$

    $2 = -2(-1)$, $2 = 2$,

    $2 = -1 - (-1)$, $2 = 2$

    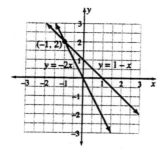

5. In $y = 3$, $m = 0$, $b = 3$

    In $x = -1$, m is undefined, no y-intercept

    Intersection is at $(-1, 3)$

    $3 = 3$,

    $-1 = -1$

    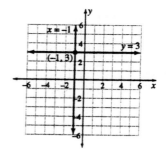

3. In $y = x$, $m = 1$, $b = 0$

    In $y = 2x + 3$, $m = 2$, $b = 3$

    Intersection is at $(-3, -3)$

    $-3 = -3$,

    $-3 = 2(-3) + 3$, $-3 = -3$

    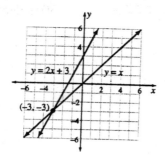

7. Solving for y; $y = -x + 5$, $m = -1$, $b = 5$

    Both equations describe the same graph, coincident lines with an infinite number of solutions.

    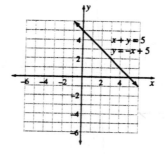

Section 7.1 (con't)

9. Solving for y, y = x - 3, m = 1, b = -3

    In y = x - 6, m = 1, b = -6

    No point of intersection, same slope different y-intercepts, parallel lines.

    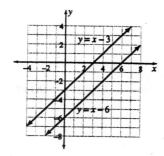

11. In y = x - 4, m = 1, b = -4

    Solving for y, y = -3x + 4, m = -3, b = 4

    Intersection is at (2, -2)

    -2 = 2 - 4, -2 = -2,

    -2 = -3(2) + 4, -2 = -2

    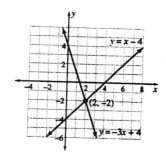

13. Solving for y, y = 2x - 1, m = 2, b = -1

    In y = 4x + 3, m = 4, b = 3

    Intersection is at (-2, -5)

    -5 = 2(-2) -1, -5 = -5,

    -5 = 4(-2) + 3, -5 = -5

    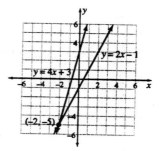

15. Solving for y, $y = -\frac{2}{3}x + 4$,

    $m = -\frac{2}{3}$, b = 4

    Solving for y, y = -x + 5, m = -1, b = 5

    Intersection is at (3, 2)

    $2 = -\frac{2}{3}(3) + 4$, 2 = 2,

    2 = -(3) + 5, 2 = 2

    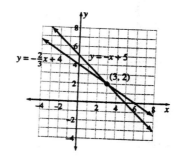

Section 7.1 (con't)

17. $2x + 2y = 100$, $2y = -2x + 100$,

$y = -x + 50$, $m = -1$, $b = 50$

In $y = 20 - x$, $m = -1$, $b = 20$

Same slope, different y-intercepts,

Parallel lines

19. $y = 55x$, $m = 55$, $b = 0$

$y = 25x$, $m = 25$, $b = 0$

Different slopes, not parallel

21. $y = 4x$, $m = 4$, $b = 0$

$y = \frac{1}{4}(12 + 16x)$, $y = 3 + 4x$, $m = 4$, $b = 3$

Same slope, different y-intercepts,

Parallel lines

23. $y + 60x = 300$, $y = -60x + 300$, $m = -60$,

$b = 300$

$y = 60 - 300x$, $m = -300$, $b = 60$

Different slopes, different y-intercepts

Not coincident

25. $y = x + 0.15x$, $y = 1.15x$, $m = 1.15$, $b = 0$

$y = 1.15x$, same equation as above,

Same slope, same y-intercept,

Coincident lines.

27. $2x + y = 10$, $y = -2x + 10$, $m = -2$, $b = 10$

$2y + x = 10$, $2y = -x + 10$, $y = -\frac{1}{2}x + 5$,

$m = -\frac{1}{2}$, $b = 5$

Different slopes, different y-intercepts

Not coincident

29. a. From the graph the y-intercept = 5,
slope is -2, equation is $y = -2x + 5$

b. From the graph the y-intercept = 3,
slope is -1, equation is $y = -x + 3$

Intersection is at (2, 1)

$1 = -2(2) + 5$, $1 = 1$, $1 = -(2) + 3$, $1 = 1$

31. a. From the graph the y-intercept = 1,
slope is $\frac{3}{2}$, equation is $y = \frac{3}{2}x + 1$

b. From the graph the y-intercept = -1,
slope is $\frac{5}{2}$, equation is $y = \frac{5}{2}x - 1$

Intersection is at (2, 4)

$4 = \frac{3}{2}(2) + 1$, $4 = 4$, $4 = \frac{5}{2}(2) - 1$, $4 = 4$

Section 7.1 (con't)

33. a. For Sense Rent-A-Car; m = $0.05,

b = $50, y = $0.05x + $50

For Herr's Rent-A-Car; m = $0.20,

b = $20, y = $0.20x + $20

b.

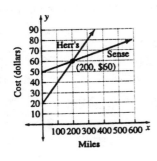

c. Intersection is at (200, $60), this is the mileage where both companies cost $60.

d. Herr's has the steeper graph because the slope is larger.

e. The y-intercept is the rental fee.

f. Sense Rent-A-Car is a better deal for distances over 200 miles, 200 < x < ∞.

35. a. y = 50 + 0.12x rewards big sales because the slope is larger.

b.

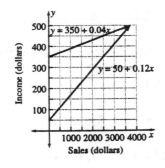

c. Intersection is at (3750, 500)

d. y = 350 + 0.04x would be preferred when sales are less than 3750.

e. Doubling the rate per dollar sales is the same as doubling the slope. New equations are: y = 350 + 0.08x and y = 50 + 0.24x

37. a. From the graph the total cost for 100 people is $525, or using the equation; C = 150 + 3.75(100), C = 525.

b. From the graph the total revenue is $500, or using the equation; R = 5(100), R = 500.

c. Profit is R - C, 500 - 525 = -25

d. The cost graph is on top when total registration is less than the breakeven point.

Section 7.1 (con't)

39. a.   C = $8.50x + $250 - $200,

C = $8.50x + $50

b.   R = $10x

c.

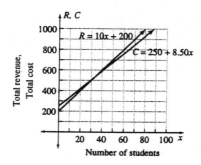

d.   From the graph the breakeven point is ≈ (34, 340).

$340 ≈ 8.50(34) + 50$, $340 ≈ 339$,

$340 = 10(34)$

The actual breakeven point is between 33 and 34, since we must have whole people we take the larger number.

41. a.

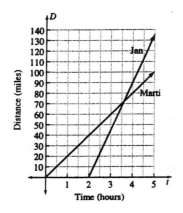

b.   Marti's distance is D = 20t

c.   Jan starts 2 hours after 7 a.m., her distance is D = 45(t - 2) = 45t - 90

d.   Intersection is at (3.6, 72)

e.   The intersection is the elapsed time and distance where Jan catches up to Marti.

f.   Jan's graph is steeper, the slope (mph) is larger.

g.   Jan starts 2 hours after Marti.

Section 7.2

1.

Item	Quantity	Value $	QV $
Nickels	x	0.05	0.05x
Quarters	y	0.25	0.25y
Total	22		2.90

$x + y = 22$, $0.05x + 0.25y = 2.90$

3.

Item	Quantity	Value $	QV $
Dimes	x	0.10	0.10x
Quarters	y	0.25	0.25y
Total	26		5.45

$x + y = 26$, $0.10x + 0.25y = 5.45$

5.

Item	Quantity	Value $	QV $
Nickels	x	0.05	0.05x
Dimes	y	0.10	0.10y
Total	65		5.40

$x + y = 65$, $0.05x + 0.10y = 5.40$

7.

Job	Hours	Wage $	Earnings $
A	x	5.75	5.75x
B	y	8.50	8.50y
Total	43		316

$x + y = 43$, $5.75x + 8.50y = 316$

9.

Grocery lbs	Restaurant lbs	Total lbs
x	y	5250

$x + y = 5250$, $x = 9y$

For exercises 11 to 17, guesses will vary.

11.

Long seg.	Short seg.	Total (m)
1 + 5 = 6	1	6 + 1 = 7
2 + 5 = 7	2	7 + 2 = 9
3 + 5 = 8	3	8 + 3 = 11
x	y	10

$x + y = 10$, $x = y + 5$

13.

Project A hr	Project B hr	Total hr
10	10 + 28 = 38	10 + 38 = 48
20	20 + 28 = 48	20 + 48 = 68
50	50 + 28 = 78	50 + 78 = 128
x	y	176

$x + y = 176$, $y = x + 28$

15.

Persons	Fun Base $	Papa's $
8	30	3(8) = 24
10	30 + 3.5(2) = 37	3(10) = 30
12	30 + 3.5(4) = 44	2(12) = 36
for x > 8	30 + 3.50(x - 8)	3x
for x ≤ 8	30	3x

For Fun Base:

if $x \leq 8$, $y = 30$,

if $x > 8$, $y = 30 + 3.5(x - 8)$

For Papa's: $y = 3x$

Section 7.2 (con't)

17.

Persons	Gymnastics $	Farrell's $
10	70	3.95(10) = 39.50
15	70	3.95(15) = 59.25
20	70 + 3(5) = 85	3.95(20) = 79.00
for x ≤ 15	70	3.95x
for x > 15	70 + 3(x - 15)	3.95x

For American Gymnastics:

if x ≤ 15, y = 70

if x > 15, y = 70 + 3(x - 15)

For Farrell's: y = 3.95x

19.

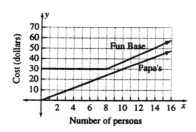

Papa's will always be less.

21.

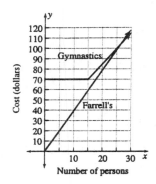

Cost is equal at ≈ (27, 104)

23. Let w = cost of watch, l = cost of locket, c = cost of chain:

w + l + c = \$620, l = c + 20, w = 2l

The watch cost is \$320, the locket cost is \$160 and the chain cost is \$140.

25. Let n = notebooks, p = paperback books and b = hard bound books:

n + p + b = 20, p = b + 3, n = p - 1

Andre purchased 7 notebooks, 8 paperback books and 5 hard bound books.

27. a. One variable requires one equation.

   b. Two variables require two equations.

   c. Three variables require three equations.

Section 7.3

1. $L = 2W$,

$$W = \frac{L}{2}$$

3. $a + b = c$,

$b = c - a$

5. $C = 2\pi r$,

$$r = \frac{C}{2\pi}$$

7. $x - y = 5$,

$x = 5 + y$,

$y = x - 5$

9. $C = \pi d$,

$$d = \frac{C}{\pi}$$

11. $2x + 3(2) = 12$,

$2x + 6 = 12$,

$2x = 6$,

$x = 3$

13. $-5x - 6(-3) = 3$,

$-5x + 18 = 3$

$18 = 3 + 5x$

$15 = 5x$

$x = 3$

15. y, $3x + y = 4$,

$y = 4 - 3x$

17. x, $x - 4y = 5$,

$x = 5 + 4y$

19. x, $5y - x = 9$,

$5y = 9 + x$,

$x = 5y - 9$

21. y, $3x - y = -2$,

$3x = -2 + y$,

$y = 3x + 2$

23. Substitute $y = x - 8$ into $3x + y = 4$,

$3x + (x - 8) = 4$,

$4x - 8 = 4$,

$4x = 12$,

$x = 3$

Now substitute $x = 3$ into $y = x - 8$,

$y = 3 - 8$,

$y = -5$

Check your solutions:

$-5 = 3 - 8$?

$3(3) + (-5) = 4$?

Section 7.3 (con't)

25. Substitute $y = 5x + 5$ into $y - 3x = 9$,

$(5x + 5) - 3x = 9$,

$2x + 5 = 9$,

$2x = 4$,

$x = 2$

Now substitute $x = 2$ into $y = 5x + 5$,

$y = 5(2) + 5$,

$y = 15$

Check your solutions:

$15 = 5(2) + 5$?

$15 - 3(2) = 9$?

27. Substitute $x = 9 + 5y$ into $4x + 5y = 11$,

$4(9 + 5y) + 5y = 11$,

$36 + 20y + 5y = 11$,

$36 + 25y = 11$,

$25y = -25$,

$y = -1$

Now substitute $y = -1$ into $x = 9 + 5y$,

$x = 9 + 5(-1)$,

$x = 4$

Check your solutions:

$4 = 9 + 5(-1)$?

$4(4) + 5(-1) = 11$?

29. First solve $2x - y = 1$ for y,

$2x - y = 1$,

$2x = 1 + y$,

$y = 2x - 1$

Substitute for y in $2y = 3x + 3$,

$2(2x - 1) = 3x + 3$,

$4x - 2 = 3x + 3$,

$x - 2 = 3$,

$x = 5$

Now substitute $x = 5$ into $y = 2x - 1$,

$y = 2(5) - 1$,

$y = 9$

Check your solutions:

$2(5) - 9 = 1$?

$2(9) = 3(5) + 3$?

31. Solve $3x + y = 7$ for y,

$y = 7 - 3x$

Substitute into $2x + 3y = 0$,

$2x + 3(7 - 3x) = 0$,

$2x + 21 - 9x = 0$,

$-7x = -21$,

$x = 3$

solution continued on next page

Section 7.3 (con't)

31. Substitute into $y = 7 - 3x$,

$y = 7 - 3(3)$,

$y = -2$

Check your solutions,

$2(3) + 3(-2) = 0$?

$3(3) + (-2) = 7$?

33. Both equations are solved for y, set them equal to each other,

$\frac{4}{3}x = -\frac{8}{3}x + 8$,

$4x = -8x + 24$,

$12x = 24$,

$x = 2$

Substitute into $y = \frac{4}{3}x$,

$y = \frac{4}{3}(2)$,

$y = \frac{8}{3}$,

Check your solutions,

$\frac{8}{3} = \frac{4}{3}(2)$?
$\frac{8}{3} = -\frac{8}{3}(2) + 8$?

35. Substitute $y = 3x + 4$ into $3x - y = 8$,

$3x - (3x + 4) = 8$,

$3x - 3x - 4 = 8$,

$-4 = 8$, this is a false statement, the system of equations has no solution

37. Substitute $y = 2x - 3$ into $y - 2x = 5$,

$(2x - 3) - 2x = 5$,

$2x - 3 - 2x = 5$,

$-3 = 5$, this is a false statement, the system of equations has no solution

39. Substitute $x = 7 - y$ into $x + y = 7$,

$(7 - y) + y = 7$.

$7 = 7$, this is always true, the system of equations has an infinite number of solutions

41. Set equations equal to each other,

$3x + 2 = -x - 2$,

$4x + 2 = -2$,

$4x = -4$,

$x = -1$

Substitute into one of the equations,

$y = -(-1) - 2$,

$y = 1 - 2$,

$y = -1$

Check your solutions,

$-1 = 3(-1) + 2$?

$-1 = -(-1) - 2$?

Section 7.3 (con't)

43. Set equations equal to each other,

$-x - 2 = 3x - 2,$

$-2 = 4x - 2,$

$0 = 4x,$

$x = 0$

Substitute into one of the equations,

$y = -(0) - 2.$

$y = -2$

Check your solutions,

$-2 = -(0) - 2?$

$-2 = 3(0) - 2?$

45. Solve $2y - x = -10$ for x, $x = 2y + 10$

Substitute into $2x + 3y = 6,$

$2(2y + 10) + 3y = 6,$

$4y + 20 + 3y = 6,$

$7y = -14,$

$y = -2$

Substitute into $x = 2y + 10,$

$x = 2(-2) + 10,$

$x = 6$

Check your solutions,

$2(-2) - (6) = -10?$

$2(6) + 3(-2) = 6?$

47. Substitute $y = \frac{1}{2} x - 2$ into $2y + 4 = x,$

$2(\frac{1}{2} x - 2) + 4 = x,$

$x - 4 + 4 = x,$

$x = x$, this is always true, system of equations has an infinite number of solutions

49. Solve $x + y = 50$ for y, $y = 50 - x,$

Substitute into $7.50x + 9.75y = 435.75,$

$7.50x + 9.75(50 - x) = 435.75,$

$7.50x + 487.50 - 9.75x = 435.75,$

$487.50 - 2.25x = 437.75,$

$51.75 = 2.25x,$

$x = 23$

Substitute into $y = 50 - x,$

$y = 50 - 23,$

$y = 27$

Check your solutions,

$23 + 27 = 50?$

$7.50(23) + 9.75(27) = 435.75?$

Section 7.3 (con't)

51. Set equations equal to each other,

$45 + 5(x - 10) = 30 + 3x,$

$45 + 5x - 50 = 30 + 3x,$

$5x - 5 = 30 + 3x,$

$2x = 35,$

$x = 17.5$

Substitute into one of the equations,

$y = 30 + 3(17.5),$

$y = 82.5$

Check your solutions,

$82.5 = 45 + 5(17.5 - 10)?$

$82.5 = 30 + 3(17.5)?$

The solution check step is not shown in the remaining exercises. QV tables may be used to set up the system of equations.

53. Let x and y represent the 2 pieces of ribbon, with y being the longer piece.

$x + y = 20, \quad y = x + 3$

$x + (x + 3) = 20,$

$2x + 3 = 20,$

$2x = 17,$

$x = 8.5$

$y = 8.5 + 3, \quad y = 11.5$

The ribbons are 8.5 yd and 11.5 yd long.

55. Let q be the number of quarters and n be the number of nickels.

$0.25q + 0.05n = 4.60, \quad q + n = 24$

$q = 24 - n$

$0.25(24 - n) + 0.05n = 4.60$

$6 - 0.25n + 0.05n = 4.60$

$6 = 4.60 + 0.20n$

$1.40 = 0.20n$

$n = 7$

$q = 24 - 7, \quad q = 17$

Yoko has 17 quarters and 7 nickels

57. Let d be the number of dimes and q be the number of quarters.

$d + q = 28, \quad 0.10d + 0.25q = 4.45$

$d = 28 - q,$

$0.10(28 - q) + 0.25q = 4.45,$

$2.80 - 0.10q + 0.25q = 4.45,$

$0.15q = 1.65,$

$q = 11$

$d = 28 - 11, \quad d = 17$

Chen Chen has 17 dimes and 11 quarters

Section 7.3 (con't)

59. Let h = height and w = width.

Recall perimeter is 2h + 2w

$40 = 2h + 2w, \; h = 4 + w,$

$40 = 2(4 + w) + 2w,$

$40 = 8 + 2w + 2w$

$32 = 4w,$

$w = 8$

$h = 4 + 8, \; h = 12$

Height is 12 in, width is 8 in

61. Let h = height and w = width.

$58 = 2h + 2w, \; h = 2w - 1$

$58 = 2(2w - 1) + 2w,$

$58 = 4w - 2 + 2w,$

$60 = 6w,$

$w = 10$

$h = 2(10) - 1, \; h = 19$

Height is 19 cm, width is 10 cm

63. Let p = grams of peanuts and c = grams of cashews.

$p + c = 270, \; 0.27p + 0.16c = 66$

$c = 270 - p,$

$0.27p + 0.16(270 - p) = 66$

$0.27p + 43.2 - 0.16p = 66$

$0.11p = 22.8$

$p \approx 207$

$c = 270 - 207, \; c = 63$

Mixture contains approximately 207 g of peanuts and 63 g of cashews.

65. a. $A + B = 180, \; A = B + 24$

$(B + 24) + B = 180,$

$2B = 156,$

$B = 78$

$A = 78 + 24, \; A = 102$

Angle A is 102° and angle B is 78°

b. $C + D = 180, \; D = C + 26$

$C + (C + 26) = 180,$

$2C = 154,$

$C = 77$

$D = 77 + 26, \; D = 103$

Angle C is 77° and angle D is 103°

Section 7.3 (con't)

65. c. $E + F = 90, \quad E = F + 2$

$(F + 2) + F = 90,$

$2F = 88,$

$F = 44$

$E = 44 + 2, \quad E = 46$

Angle E is 46° and angle F is 44°

d. $G + H = 90, \quad H = 2G + 3$

$G + (2G + 3) = 90,$

$3G = 87,$

$G = 29$

$H = 2(29) + 3, \quad H = 61$

Angle G is 29° and angle H is 61°

e. $I + J = 180, \quad J = 2I - 45,$

$I + (2I - 45) = 180,$

$3I = 225,$

$I = 75$

$J = 2(75) - 45, \quad J = 105$

Angle I is 75° and angle J is 105°

67. a. Let W = width and L = length.

$\dfrac{W}{L} = \dfrac{1}{7}, \quad 32 = 2W + 2L,$

$L = 7W,$

$32 = 2W + 2(7W),$

$32 = 16W,$

$W = 2$

$L = 7(2), \quad L = 14$

Width is 2 m and length is 14 m.

b. $\dfrac{L}{W} = \dfrac{5}{3}, \quad 320 = 2W + 2L,$

$L = \dfrac{5W}{3},$

$320 = 2W + 2\left(\dfrac{5W}{3}\right),$

$320 = 2W + \dfrac{10W}{3},$

$320 = \dfrac{16W}{3},$

$960 = 16W,.$

$W = 60$

$L = \dfrac{5(60)}{3}, \quad L = 100$

Length = 100 yd and width = 60 yd.

Section 7.3 (con't)

69. Let w = cost of the watch, k = cost of the locket and c = cost of the chain,

w + k + c = 620, k = c + 20, w = 2k,

c = k - 20,

(2k) + k + (k - 20) = 620,

4k - 20 = 620,

4k = 640,

k = 160

c = 160 - 20, c = 140

w = 2(160), w = 320

Marielena paid \$320 for the watch, \$160 for the locket and \$140 for the chain.

71. Let n = the number of notebooks, p = the number of paperbacks and h = the number of hard bound books.

n + p + h = 20, p = h + 3, n = p - 1,

h = p - 3,

(p - 1) + p + (p - 3) = 20,

3p - 4 = 20,

3p = 24,

p = 8

n = 8 - 1, n = 7

h = 8 - 3, h = 5

Andre bought 7 notebooks, 8 paperbacks and 5 hardbound books.

73. A + B + C = 180, A = 2B, B = 3C

$$C = \frac{B}{3},$$

$$(2B) + B + \frac{B}{3} = 180,$$

$$\frac{10B}{3} = 180,$$

$$B = \frac{3 \cdot 180}{10}, B = 54$$

54 = 3C,

C = 18

A = 2(54), A = 108

A = 108, B = 54, C = 18

75. A = 2B, C = A + 20, A + B + C = 180

$$B = \frac{A}{2},$$

$$A + \frac{A}{2} + (A + 20) = 180,$$

$$\frac{5A}{2} = 160,$$

$$A = \frac{2 \cdot 160}{5}, A = 64$$

$$B = \frac{64}{2}, B = 32$$

C = 64 + 20, C = 84

Angle A = 64°, angle B = 32° and angle C = 84°.

Section 7.3 (con't)

77. A = B, C = 2A, A + B + C = 180

A + A + 2A = 180,

4A = 180,

A = 45

B = 45

C = 2(45), C = 90

The equal angles are each 45° and the third angle is 90°.

79. $l = w + 7$, w = 4h, $l + w + h = 34$

$h = \dfrac{w}{4}$,

$(w + 7) + w + \dfrac{w}{4} = 34$,

$\dfrac{9w}{4} = 27$,

$w = \dfrac{4 \cdot 27}{9}$, w = 12

$l = 12 + 7, l = 19$

$h = \dfrac{12}{4}, h = 3$

Length is 19 in, width is 12 in and height is 3 in.

81. Replace one variable with its' equivalent equation in terms of the other variable. Solve for that variable and use the value to find the other variable.

83. Choose guess and check for the simplest systems and graphing for the most complicated systems. Substitution can be used for all systems and is generally used to check answers.

Mid-Chapter 7 Test

1. $2x - y = 5000$,

   $2x = 5000 + y$,

   $y = 2x - 5000$

2. $V = \dfrac{\pi r^2 h}{3}$,

   $3V = \pi r^2 h$,

   $h = \dfrac{3V}{\pi r^2}$

3. $\dfrac{l}{w} = \dfrac{13}{6}$,

   $l = \dfrac{13w}{6}$

4. $x + y = 5000$, $3x - 2y = -2500$,

   $y = 5000 - x$,

   $3x - 2(5000 - x) = -2500$,

   $3x - 10,000 + 2x = -2500$,

   $5x = 7500$,

   $x = 1500$

   $y = 5000 - 1500$, $y = 3500$

5. $x + y = 5000$, $3x - 2y = 2500$,

   $y = 5000 - x$,

   $3x - 2(5000 - x) = 2500$,

   $3x - 10,000 + 2x = 2500$,

   $5x = 12,500$,

   $x = 2500$,

   $y = 5000 - 2500$, $y = 2500$

6. **a.** It is the intersection and therefore the solution of $y = 6 - \frac{2}{3}x$ and $y = x + 1$.

   b. Estimates should be between 4 and 5 for x and between 5 and 6 for y.

   c. $y = 10 - x$, $y = x + 1$

   $10 - x = x + 1$,

   $10 = 2x + 1$,

   $9 = 2x$,

   $x = 4.5$

   $y = 4.5 + 1$, $y = 5.5$

   Intersection is at $(4.5, 5.5)$

Mid-Chapter 7 Test (con't)

6. d. $y = 10 - x, y = 6 - \frac{2}{3}x$

$10 - x = 6 - \frac{2}{3}x,$

$10 = 6 + \frac{1}{3}x,$

$4 = \frac{1}{3}x,$

$x = 12$

$y = 10 - 12, y = -2$

Intersection is at $(12, -2)$

7. a.

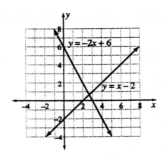

b. Estimates should be between 2 and 3 for x and between 0 and 1 for y.

c. $2x + y = 6, y = x - 2$

$2x + (x - 2) = 6,$

$3x - 2 = 6,$

$3x = 8,$

$x = \frac{8}{3}$

$y = \frac{8}{3} - 2, y = \frac{8-6}{3}, y = \frac{2}{3}$

Intersection is at $\left(\frac{8}{3}, \frac{2}{3}\right)$

8. $2x + y = 4, y = 4 - 2x,$

$2x + (4 - 2x) = 4,$

$2x + 4 - 2x = 4,$

$4 = 4,$ Always true, the system of equations has an infinite number of solutions.

9. Let x = turtle eggs and y = ostrich eggs,

$x = 12y, x + y = 195$

$12y + y = 195,$

$13y = 195,$

$y = 15$

$x = 12(15), x = 180$

!80 green turtle eggs and 15 ostrich eggs

10. Let x = 5 pound bags and y = 10 pound bags.

$x = 3y, 5x + 10y = 10,000$

$5(3y) + 10y = 10,000,$

$15y + 10y = 10,000,$

$25y = 10,000,$

$y = 400$

$x = 3(400), x = 1200$

1200 5-pound bags and 400 10-pound bags.

Mid-Chapter 7 Test (con't)

11. Let L = length and W = width,

$$2L + 2W = 114, \quad \frac{L}{W} = \frac{13}{6}, \quad L = \frac{13W}{6},$$

$$2\left(\frac{13W}{6}\right) + 2W = 114,$$

$$\frac{13W}{3} + \frac{6W}{3} = 114,$$

$$\frac{19W}{3} = 114,$$

$$W = \frac{114 \cdot 3}{19}, \quad W = 18$$

$$L = \frac{13(18)}{6}, \quad L = 39$$

Length = 39 ft, width = 18 ft

12. The intersection is where the same input gives the same output in each equation.

Section 7.4

1. a.

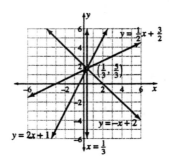

b. $y = 2x + 1$

$\underline{+\ y = -x + 2}$

$2y = x + 3$

$y = \dfrac{x}{2} + \dfrac{3}{2}$

Equation is on graph above.

c. $y = 2x + 1$

$\underline{-\ y = -x + 2}$

$0 = 3x - 1$

$3x = 1$

$x = \dfrac{1}{3}$,

Equation is on graph above.

d. All graphs intersect at the same point.

3. $x + y = -2$

$\underline{+\ x - y = 8}$

$2x + 0 = 6$

$x = 3$

$3 + y = -2,$

$y = -5$

5. $m + n = 3$

$\underline{+\ -m + n = -11}$

$0 + 2n = -8$

$n = -4$

$m + (-4) = 3,$

$m = 7$

7. Multiply the second equation by -2

$(-2)(x + 2y) = (4)(-2)$

$-2x - 4y = -8$, now add to first equation

$\underline{+\ 2x + y = -1}$

$0 - 3y = -9$

$y = 3$

$x + 2(3) = 4,$

$x = -2$

Section 7.4 (con't)

9. Multiply the first equation by -3

$(-3)(2a + b) = (-5)(-3)$

$-6a - 3b = 15$, now add 2nd equation

$\underline{+\ a + 3b = 35}$

$-5a + 0 = 50$

$a = -10$

$2(-10) + b = -5$

$b = 15$

11. Multiply the first equation by 3 and the second equation by 2, then subtract.

$(3)(2x + 3y) = (3)(3),$

$6x + 9y = 9$

$(2)(3x - 4y) = (-21)(2),$

$6x - 8y = -42$

$6x + 9y = 9$

$\underline{-\ (6x - 8y = -42)}$

$0 + 17y = 51$

$y = 3$

$2x + 3(3) = 3,$

$2x = -6$

$x = -3$

13. Multiply the first equation by 3 and the second equation by 2, then add.

$(3)(5p - 2q) = (-6)(3)$

$15p - 6q = -18$

$(2)(2p + 3q) = (9)(2)$

$4p + 6q = 18$

$\underline{+\ 15p - 6q = -18}$

$19p + 0 = 0$

$p = 0$

$5(0) - 2q = -6,$

$-2q = -6,$

$q = 3$

15. $x + y = 6$

$\underline{-\ (x + y = 10)}$

$0 + 0 = -4$, false statement, no solution

17. $(2)(x - y) = 7(2)$

$2x - 2y = 14$

$\underline{-\ (2x - 2y = 14)}$

$0 = 0$, always true, infinite number of solutions

Section 7.4 (con't)

19. $x + y = 5$

$\underline{+ \quad y - x = -13}$

$2y + 0 = -8,$

$y = -4$

$x + (-4) = 5$

$x = 9$

21. $(3)(2x + 3y) = 0(3)$

$6x + 9y = 0$

$(-2)(3x + 2y) = (5)(-2)$

$-6x - 4y = -10$

$\underline{+ \quad 6x + 9y = 0}$

$0 + 5y = -10$

$y = -2$

$2x + 3(-2) = 0$

$2x = 6$

$x = 3$

23. $7 = 5m + b$

$\underline{- (3 = 3m + b)}$

$4 = 2m + 0$

$m = 2$

$7 = 5(2) + b$

$b = -3$

25. $(-2)(0.2x + 0.6y) = (2.2)(-2)$

$-0.4x - 1.2y = -4.4$

$\underline{+ \quad 0.4x - 0.2y = 1.6}$

$0 - 1.4y = -2.8$

$y = 2$

$0.4x - 0.2(2) = 1.6$

$0.4x = 2.0$

$x = 5$

27. $(2)(0.5x + 0.2y) = (1.8)(2)$

$1.0x + 0.4y = 3.6$

$(5)(0.2x - 0.3y) = (-0.8)(5)$

$1.0x - 1.5y = -4.0$

$- (1.0x + 0.4y = 3.6)$

$0 - 1.9y = -7.6$

$y = 4$

$0.5x + 0.2(4) = 1.8$

$0.5x = 1.0$

$x = 2$

Section 7.4 (con't)

29. Let x and y be two numbers such that:

$x + y = 25$ and $x - y = 8$.

$x + y = 25$

$- (x - y = 8)$

$0 + 2y = 17$

$y = 8.5$

$x + 8.5 = 25,$

$x = 16.5$

31. Let x and y be two numbers such that:

$x + y = 20$ and $2y - 2x = 21$.

$(2)(x + y) = (20)(2)$

$2x + 2y = 40$

$+ 2y - 2x = 21$

$4y = 61$

$y = 15.25$

$x + 15.25 = 20,$

$x = 4.75$

33. $(-4)(3p + 4f) = (48)(-4)$

$-12p - 16f = -192$

$(3)(4p + 3f) = (43)(3)$

$12p + 9f = 129$

$+ \ -12p - 16f = -192$

$0 - 7f = -63$

$f = 9$

$3p + 4(9) = 48$

$3p = 12$

$p = 4$

35. $x + y = 243, \ y = x - 17$

$x + (x - 17) = 243$

$2x = 260$

$x = 130$

$y = 130 - 17, y = 113$

Section 7.4 (con't)

37. a. Let A and B be two angles.

A + B = 90° (complementary)

B = A + 50°, B - A = 50°

A + B = 90

$\underline{+\ B - A = 50}$

2B = 140

B = 70°

A + 70 = 90

A = 20°

b. Let C & D be two angles.

C + D = 180° (supplementary)

$\underline{D - C = 50°}$

2D = 230

D = 115°

C + 115 = 180,

C = 65°

39. a. A + B = 180, A = B + 24

(B + 24) + B = 180,

2B = 156,

B = 78°

A = 78 + 24, A = 102°

39. b. C + D = 180, D = C + 26

C + (C + 26) = 180,

2C = 154,

C = 77°

D = 77 + 26, D = 103°

c. E + F = 90, E = F + 2

(F + 2) + F = 90,

2F = 88,

F = 44°

E = 44 + 2, E = 46°

d. G + H = 90, H = 2G + 3

G + (2G + 3) = 90,

3G = 87,

G = 29°

H = 2(29) + 3, H = 61°

e. I + J = 180, J = 2I - 45

I + (2I - 45) = 180,

3I = 225,

I = 75°

75 + J = 180,

J = 105°

Section 7.4 (con't)

41. Let c = sugar cookies calories and
g = ginger snap calories.

$4c + 2g = 296$, $3c + 10g = 329$

$(-5)(4c + 2g) = (296)(-5)$

$-20c - 10g = -1480$

$\underline{+\quad 3c + 10g = 329}$

$-17c = -1151$,

$c \approx 67.7$

$4(67.7) + 2g = 296$,

$2g = 25.2$

$g = 12.6$

Sugar cookies have 67.7 calories and
ginger snaps have 12.6 calories.

43. Let c = cherries and g = grapes.

$15c + 22g = 126$, $20c + 11g = 113$

$(-2)(20c + 11g) = (113)(-2)$,

$-40c - 22g = -226$

$\underline{+15c + 22g = 126}$

$-25c = -100$,

$c = 4$

$15(4) + 22g = 126$,

$22g = 66$,

$g = 3$

Cherries have 4 cal. and grapes have 3 cal.

45. Let g = green olives and r = ripe olives.

$8g + 5r = 285$, $4g + 10r = 330$

$(-2)(8g + 5r) = (285)(-2)$,

$-16g - 10r = -570$

$\underline{+\quad 4g + 10r = 330}$

$-12g = -240$,

$g = 20$

$8(20) + 5r = 285$,

$5r = 125$,

$r = 25$

Green olives have 20 cal. and ripe olives
have 25 cal.

47. Let a = adult tickets and s = student
tickets.

$6a + 3s = 58.50$, $5a + 4s = 54$

$3s = 58.50 - 6a$,

$s = 19.50 - 2a$

$5a + 4(19.50 - 2a) = 54$,

$5a + 78 - 8a = 54$

$-3a = -24$

$a = 8$

$s = 19.50 - 2(8)$, $s = 3.50$

Adult tickets cost $8.00 and student
tickets cost $3.50.

Section 7.4 (con't)

49. Let s = shirts and t = ties.

$3s + 2t = 109.95, \quad 4s + t = 119.95$

$t = 119.95 - 4s$

$3s + 2(119.95 - 4s) = 109.95,$

$3s + 239.90 - 8s - 109.95,$

$-5s = -129.95,$

$s = 25.99$

$t = 119.95 - 4(25.99), \quad t = 15.99$

Shirts cost $25.99 and ties cost $15.99.

51. Let d = CD's and g = game disks.

$2d + 3g = 137.95, \quad 4d + g = 75.95$

$g = 75.95 - 4d,$

$2d + 3(75.95 - 4d) = 137.95,$

$2d + 227.85 - 12d = 137.95,$

$-10d = -89.90,$

$d = 8.99$

$g = 75.95 - 4(8.99), \quad g = 39.99$

CD's cost $8.99 and game disks cost $39.99.

53. Let g = time on gravel roads and p = time on paved roads.

$g + p = 7, \quad 40g + 65p = 405$

$g = 7 - p,$

$40(7 - p) + 65p = 405,$

$280 - 40p + 65p = 405,$

$25p = 125,$

$p = 5$

$g = 7 - 5, \quad g = 2$

Ned drove 2 hours on gravel roads and 5 hours on paved roads.

55. Let p = the speed of the plane in still air and w = the speed of the wind.

$2(p + w) = 1100, \quad 4(p - w) = 1100$

$p + w = 550,$

$p = 550 - w,$

$4[(550 - w) - w] = 1100$

$550 - 2w = 275$

$-2w = -275$

$w = 137.5$

$p = 550 - 137.5, \quad p = 412.5$

The wind speed is 137.5 mph and the speed of the plane is 412.5 mph.

Section 7.4 (con't)

57. Let r = the speed of the plane and
w = the speed of the wind.

$4.5(r - w) = 360, \ 3(r + w) = 360$

$r - w = 80,$

$r = 80 + w$

$3[(80 + w) + w] = 360,$

$80 + 2w = 120,$

$2w = 40,$

$w = 20$

$r = 80 + 20, r = 100$

The speed of the plane is 100 mph and
the speed of the wind is 20 mph.

59. Let b = the speed of the boat and c = the
speed of the current.

$5(b - c) = 20, \ 2(b + c) = 20$

$b - c = 4,$

$b = 4 + c,$

$2[(4 + c) + c] = 20,$

$4 + 2c = 10,$

$2c = 6,$

$c = 3$

$b = 4 + 3, b = 7$

The speed of the boat is 7 mph and the
speed of the current is 3 mph.

61. When walking you have the advantage
of friction between the bottom of your
feet and the ground.

63. Add or subtract equations to remove one
variable. Solve for the remaining
variable then substitute your solution
into one of the original equations and
solve for the other variable. Check your
solutions in each equation.

65. Multiply $ax + by = c$ by e on both sides
and multiply $dx + ey = f$ by b on both
sides. Subtract the equations.

Section 7.5

1. a. $0 \leq 0 + 2$, true

   $0 \geq 0 - 2$, true

   $(0, 0)$ is a solution

   b. $1 \leq 1 + 2$, true

   $1 \geq 1 - 2$, true

   $(1, 1)$ is a solution

   c. $-2 \leq 1 + 2$, true

   $-2 \geq 1 - 2$, false

   $(1, -2)$ is not a solution

   d. $3 \leq -3 + 2$, false

   $(-3, 3)$ is not a solution

3. a. $0 + 0 > 4$, false

   $(0, 0)$ is not a solution

   b. $2 + 2 > 4$, false

   $(2, 2)$ is not a solution

   c. $5 + 0 > 4$, true

   $5 - 0 > 4$, true

   $(5, 0)$ is a solution

   d. $0 + 5 > 4$, true

   $0 - 5 > 4$, false

   $(0, 5)$ is not a solution

5.

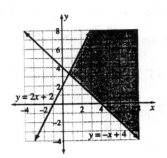

7.

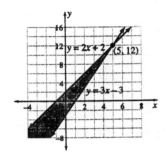

9.

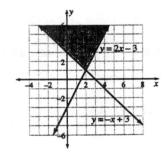

11.

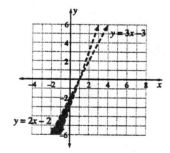

Section 7.5 (con't)

13.

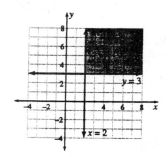

15.

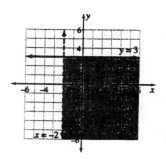

17.

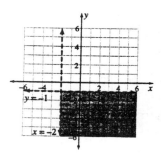

19.

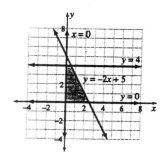

21.

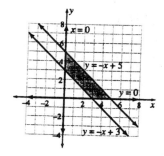

23.

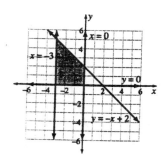

25. $x + y \le 12, \; x \le 4, \; x \ge 0, \; y \ge 0$

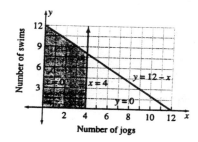

27. $15x + 8y \le 240, \; y \le 12, \; x \ge 0, \; y \ge 0$

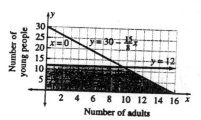

29. $x + y \le 45{,}000, \; y \le 5000, \; x \ge 0, \; y \ge 0$

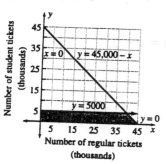

Chapter 7 Review

1. $y - 2x = 5000,$

$y = 5000 + 2x$

3. $C = 2\pi r,$

$r = \dfrac{C}{2\pi}$

5. $\dfrac{a}{b} = \dfrac{5}{8},$

$5b = 8a,$

$b = \dfrac{8a}{5}$

7.

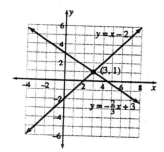

9.

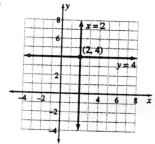

11. $y = x + 2, \ y = 3x + 3$

$3x + 3 = x + 2,$

$2x + 3 = 2,$

$2x = -1,$

$x = -\frac{1}{2}, \ y = -\frac{1}{2} + 2, y = \frac{3}{2}$

13. $x + y = 5000, \ 3x - 2y = 12,500$

$x = 5000 - y$

$3(5000 - y) - 2y = 12,500,$

$15,000 - 3y - 2y = 12,500,$

$-5y = -2500,$

$y = 500$

$x = 5000 - 500, \ x = 4500$

15. $3 = -2m + b$

$\underline{- (2 = 3m + b)}$

$1 = -5m,$

$m = -\frac{1}{5}$

$3 = -2(-\frac{1}{5}) + b,$

$3 - \frac{2}{5} = b,$

$b = \frac{15}{5} - \frac{2}{5}, \ b = \frac{13}{5}$

17. $2x - 3y = 7$

$\underline{+ \ 4x + 3y = -1}$

$6x = 6,$

$x = 1$

$2(1) - 3y = 7$

$-3y = 5,$

$y = -\frac{5}{3}$

Chapter 7 Review (con't)

19. $(2)(5x + 3y) = (-18)(2)$

$10x + 6y = -36$

$(3)(3x - 2y) = (-7)(3)$

$9x - 6y = -21$

$+\ \underline{10x + 6y = -36}$

$19x = -57,$

$x = -3$

$3(-3) - 2y = -7,$

$-2y = 2,$

$y = -1$

21. $x + y = 15,\ \ y = -6x$

$x + (-6x) = 15,$

$-5x = 15,$

$x = -3$

$y = -6(-3),\ y = 18$

23. $(2)(3y - 4x) = (2)(2)$

$6y - 8x = 4$

$8x = 6y + 4$

$8x - 6y = 4$

$+\ \underline{6y - 8x = 4}$

$0 = 4$, false statement, no solution

25. $0 = 0$ is always true therefore the lines on the graph will be coincident.

27. Let c = centipede legs and m = millipede legs.

$c = m - 356,\ \ c + m = 1064$

$(m - 356) + m = 1064,$

$2m = 1420,$

$m = 710$

$c = 710 - 356,\ c = 354$

The centipede has 354 legs and the millipede has 710 legs.

29. Let m = English muffin and e = fried egg

$m + 2e = 330,\ \ 3m + e = 515$

$m = 330 - 2e,$

$3(330 - 2e) + e = 515,$

$990 - 6e + e = 515,$

$-5e = -475,$

$e = 95,$

$m = 330 - 2(95),\ m = 140$

The English muffin has 140 calories and the fried egg has 95 calories.

Chapter 7 Review (con't)

31. Let c = carbohydrate and f = fat.

$5c + 2f = 38$, $2c + 6f = 62$

$2c = 62 - 6f$,

$c = 31 - 3f$

$5(31 - 3f) + 2f = 38$,

$155 - 15f + 2f = 38$

$-13f = -117$,

$f = 9$

$5c + 2(9) = 38$,

$5c = 20$,

$c = 4$

One gram of carbohydrate has 4 calories and 1 gram of fat has 9 calories.

33. Let s = stools and t = tables

$s + t = 19$, $3s + 4t = 64$

$s = 19 - t$,

$3(19 - t) + 4t = 64$,

$57 - 3t + 4t = 64$,

$t = 7$

$s + 7 = 19$,

$s = 12$

Mr. Schaaf has 12 stools and 7 tables.

35. Let t = speed of the trout and c = speed of the current.

$0.8(t - c) = 8$, $0.8(t + c) = 16$

$0.8t - 0.8c = 8$

$+ \underline{0.8t + 0.8c = 16}$

$1.6t = 24$,

$t = 15$

$0.8(15) - 0.8c = 8$,

$-0.8c = -4$,

$c = 5$

The speed of the trout is 15 mph and the speed of the current is 5 mph.

37. Let x = one of the two equal sides and y = the third side.

$2x + y = 32$, $x = y + 2.5$,

$2(y + 2.5) + y = 32$,

$2y + 5 + y = 32$,

$3y = 27$,

$y = 9$

$x = 9 + 2.5$, $x = 11.5$

The 2 equal side are each 11.5 inches and the third side is 9 inches.

Chapter 7 Review (con't)

39. Let p = protein, f = fat and

c = carbohydrates

$4p + 9f + 4c = 198$, $p + f + c = 37$,

$p = c + 5$, substitute for p in the other 2 equations

$4(c + 5) + 9f + 4c = 198$,

$4c + 20 + 9f + 4c = 198$

$9f + 8c = 178$ (eqn. A)

$(c + 5) + f + c = 37$,

$f + 2c = 32$, solve this for f

$f = 32 - 2c$, substitute this for f in eqn. A

$9(32 - 2c) + 8c = 178$,

$288 - 18c + 8c = 178$,

$-10c = -110$,

$c = 11$, use this to find f and p

$f = 32 - 2(11)$, $f = 10$,

$p = 11 + 5$, $p = 16$,

Check solutions in all equations:

$4(16) + 9(10) + 4(11) = 198$?

$64 + 90 + 44 = 198$, true

$16 + 10 + 11 = 37$, true

$16 = 11 + 5$, true. There are 16 g of protein, 10 g of fat and 11 g of carbohydrate in this serving of shrimp.

41. Let p = protein, f = fat and

c = carbohydrates

$4p + 9f + 4c = 903$, $p + f + c = 137$,

$f = c + 44$, substitute for f in the other 2 equations.

$4p + 9(c + 44) + 4c = 903$,

$4p + 9c + 396 + 4c = 903$,

$4p + 13c = 507$ (eqn. B)

$p + (c + 44) + c = 137$,

$p + 2c = 93$, solve this for p

$p = 93 - 2c$, substitute this for p in eqn. B

$4(93 - 2c) + 13c = 507$,

$372 - 8c + 13c = 507$,

$5c = 135$,

$c = 27$, use this to find p and f

$p = 93 - 2(27)$, $p = 39$,

$f = 27 + 44$, $f = 71$

Check solutions in all equations:

$4(39) + 9(71) + 4(27) = 903$?

$156 + 639 + 108 = 903$, true

$39 + 71 + 27 = 137$, true

$71 = 27 + 44$, true

There are 39 g of protein, 71 g of fat and 27 g of carbohydrates in the peanuts.

Chapter 7 Review (con't)

43. a. j. Choose graphing when the equations are easily placed into $y = mx + b$ form.

b. h. Choose a table and graph from a graphing calculator when there is no algebraic method for a solution.

c. g. Choose guess and check when you are trying to understand a problem.

d. f. Choose substitution when one variable has a 1 or -1 coefficient.

e. i. Choose elimination when coefficients are integers or with multiplication can be made into integers.

45.

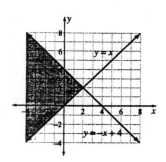

47.

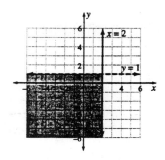

49.

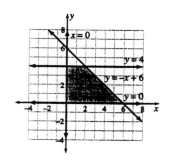

51. $55x + 65y \leq 715$, $y \leq 2$, $x \geq 0$, $y \geq 0$

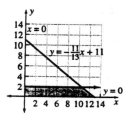

Chapter 7 Test

1. $3x - y = 400,$

   $3x = 400 + y$

   $y = 3x - 400$

2. $x - 2y = 3,$

   $x = 3 + 2y,$

   $2y = x - 3,$

   $y = \dfrac{x}{2} - \dfrac{3}{2}$

3. $a + 2b = 6,\ a = 6 - 2b$

   $3a - b = -17,$

   $3(6 - 2b) - b = -17,$

   $18 - 6b - b = -17$

   $-7b = -35,$

   $b = 5$

   $a = 6 - 2(5),\ a = -4$

4. Solve both equations for b,

   $b = 2 - 5m,\ b = -1 - m$

   $2 - 5m = -1 - m$

   $3 = 4m$

   $m = 0.75$

   $b = -1 - 0.75,\ b = -1.75$

5. Solve both equations for y,

   $y = 5 - x,\ y = -x + 5$

   $5 - x = -x + 5,$ always true, infinite number of solutions.

6. Solve both equations for y,

   $y = 2x - 1,\ y = 3 + 2x$

   $2x - 1 = 3 + 2x,$

   $0 = 4,$ always false, no solutions

7. Multiply second equation by -2,

   $(-2)(5x - 2y) = (47)(-2)$

   $-10x + 4y = -94$

   $\underline{+\quad 3x - 4y = 3}$

   $-7x = -91,$

   $x = 13$

   $5(13) - 2y = 47,$

   $-2y = -18,$

   $y = 9$

8. Solve second equation for x,

   $x = 900 + y,$

   $3(900 + y) + 2y = 5700,$

   $2700 + 5y = 5700$

   $5y = 3000,$

   $y = 600$

   $x = 900 + 600,\ x = 1500$

Chapter 7 Test (con't)

9. The graphs are coincident lines.

10. The lines are parallel.

11. (6, 2) is the point of intersection of

$y = 8 - x$ and $y = 4 - \frac{1}{3}x$.

Substituting the intersection coordinates into the equations will make both equations true.

12.

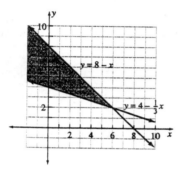

13. $1 < 2(7) + 2$, true

$1 \leq 8 - 7$, true

(7, 1) is a solution set because it makes both inequalities true.

14. Let c = caterpillar and b = butterfly

$c = b + 10$, $6c + 8b = 144$

$6(b + 10) + 8b = 144$,

$6b + 60 + 8b = 144$

$14b = 84$,

$b = 6$, $c = 6 + 10$, $c = 16$

The caterpillar has 16 legs and the butterfly has 6 legs.

15. Let p = peanuts and c = cashews

$16p + 5c = 135$, $20p + 25c = 405$

$(-5)(16p + 5c) = (135)(-5)$

$-80p - 25c = -675$

$+ \underline{20p + 25c = 405}$

$-60p = -270$

$p = 4.5$

$20(4.5) + 25c = 405$,

$25c = 315$,

$c = 12.6$

Peanuts have 4.5 calories and cashews have 12.6 calories

16. Let r = the speed of the dolphin and c = the speed of the current

$3(r + c) = 135$, $4(r - c) = 100$

$r + c = 45$

$+ \underline{r - c = 25}$

$2r = 70$,

$r = 35$

$35 + c = 45$

$c = 10$

The speed of the dolphin is 35 mph and the speed of the current is 10 mph.

Chapter 7 Test (con't)

17. $x + y \leq 216, \ y \leq 50, x \geq 0, \ y \geq 0$

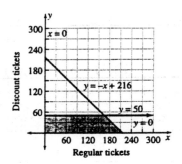

18. If a variable has a coefficient of 1 or -1 use substitution.

Section 8.1

1. $x = \sqrt{3^2 + 5^2}$

$x = \sqrt{34}$, $x \approx 5.83$

3. $x = \sqrt{9^2 + 6^2}$

$x = \sqrt{117}$, $x \approx 10.82$

5. $8^2 = x^2 + 5^2$

$x^2 = 64 - 25$

$x = \sqrt{39}$, $x \approx 6.24$

7. $7^2 + 8^2 = 9^2$?

$49 + 64 \neq 81$, not a right triangle

9. $12^2 + 16^2 = 20^2$?

$144 + 256 = 400$, right triangle

11. $7^2 + 24^2 = 25^2$?

$49 + 576 = 625$, right triangle

13. $\dfrac{x}{6} = \dfrac{28}{8}$, $x = \dfrac{6 \cdot 28}{8}$, $x = 21$

$\dfrac{w}{10} = \dfrac{28}{8}$, $w = \dfrac{10 \cdot 28}{8}$, $w = 35$

$6^2 + 8^2 = 10^2$?

$36 + 64 = 100$, right triangle

$21^2 + 28^2 = 35^2$?

$441 + 784 = 1225$, right triangle

15.

Leg	Leg	Hypotenuse
3	4	5
6	$(2)(4) = 8$	$(2)(5) = 10$
$(6)(3) = 18$	$(6)(4) = 24$	30
$(3)(3) = 9$	12	$(3)(5) = 15$
$1 = \frac{3}{3}$	$\frac{4}{3}$	$\frac{5}{3}$

17. a. $(3x)^2 = 3^2 x^2 = 9x^2$

b. $(4x)^2 = 4^2 x^2 = 16x^2$

c. $(5x)^2 = 5^2 x^2 = 25x^2$

d. $(6x)^2 = 6^2 x^2 = 36x^2$

19. $8^2 = x^2 + x^2$,

$2x^2 = 64$, $x^2 = 32$

$x = \sqrt{32}$, $x \approx 5.7$

21. $52^2 = x^2 + (5x)^2$,

$2704 = x^2 + 25x^2$, $2704 = 26x^2$,

$x^2 = 104$, $x = \sqrt{104}$, $x \approx 10.2$

23. $15^2 = x^2 + (2x)^2$,

$225 = x^2 + 4x^2$, $225 = 5x^2$,

$x^2 = 45$, $x = \sqrt{45}$, $x \approx 6.7$

25. $\sqrt{12^2 + 3^2} = \sqrt{144 + 9} = \sqrt{153} \approx 12.4$

Ladder would need to be 12.4 ft long.

27. $\sqrt{9^2 + 2.25^2} = \sqrt{81 + 5.06} \approx 9.3$

Ladder would need to be 9.3 ft long.

Section 8.1 (con't)

29. $\dfrac{14}{b} = \dfrac{4}{1}$, $b = \dfrac{14}{4}$, $b = 3.5$

Safe position is 3.5 ft from the base of the ladder to the wall.

$\sqrt{14^2 + 3.5^2} = \sqrt{196 + 12.25} \approx 14.4$

Ladder would need to be 14.4 ft long.

31. From the safe ladder ratio, if the base is x the height would be 4x. Now using the Pythagorean theorem:

$12^2 = x^2 + (4x)^2$, $144 = x^2 + 16x^2$,

$17x^2 = 144$, $x^2 = 144 \div 17$

$x = \sqrt{\frac{144}{17}}$, $x \approx 2.91$ ft

$4x \approx (4)(2.91)$, $4x \approx 11.64$ ft

$(0.91)(12) \approx 11$,

$(0.64)(12) \approx 8$

base is ≈ 2.91 ft or 2 ft 11 in

height is ≈ 11.64 ft or 11 ft 8 in

33. $18^2 = x^2 + (4x)^2$, $324 = 17x^2$,

$x^2 = 324 \div 17$,

$x = \sqrt{\frac{324}{17}}$, $x \approx 4.37$

$4x \approx (4)(4.37)$, $4x \approx 17.48$

$(0.37)(12) \approx 4$,

$(0.48)(12) \approx 6$

base is ≈ 4.37 ft or 4 ft 4 in,

height is ≈ 17.48 ft or 17 ft 6 in

35.

$\sqrt{250^2 + 30^2} =$

$\sqrt{63400} \approx 251.8$ miles

37.

$\sqrt{(3 \cdot 200)^2 + (3 \cdot 32)^2} =$

$\sqrt{600^2 + 96^2} = \sqrt{369216} \approx 607.6$ miles

39. The right triangle formed from the peak of the roof has a base of $36 \div 2 = 18$ and a hypotenuse of 20. The total height, h, is 11 + the height of the triangle.

$h = 11 + \sqrt{20^2 - 18^2}$,

$h = 11 + \sqrt{76}$, $h \approx 11 + 8.72$,

$h \approx 19.72$ ft

Section 8.1 (con't)

41. With a slope of 3 to 14 one edge of the

roof line is $\sqrt{3^2 + 14^2} = \sqrt{205} \approx 14.3$ ft.

Area of roof is $(14.3)(40)(2) \approx 1146$ ft^2

43. One edge of roof line is

$\sqrt{9^2 + 14^2} = \sqrt{277} \approx 16.6$ ft.

Area of roof is $(16.6)(40)(2) \approx 1332$ ft^2

Section 8.2

1. a. $\sqrt{81} = \sqrt{9^2} = 9$

   b. $\sqrt{1.96} = \sqrt{1.4^2} = 1.4$

   c. $\sqrt{0.04} = \sqrt{0.2^2} = 0.2$

   d. $\sqrt{3600} = \sqrt{60^2} = 60$

3. a. $8^2 = 64,\ 9^2 = 81,$

   $\sqrt{80}$ is between 8 and 9.

   b. $7^2 = 49,\ 8^2 = 64,$

   $\sqrt{54}$ is between 7 and 8

   c. $14^2 = 196,\ 15^2 = 225,$

   $\sqrt{210}$ is between 14 and 15

   d. $4^2 = 16,\ 5^2 = 25,$

   $\sqrt{18}$ is between 4 and 5

5.

Square root	Value	Rational or Irrational ?
$\sqrt{36}$	6	rational
$\sqrt{15}$	3.873	irrational
$\sqrt{25}$	5	rational
$\sqrt{35}$	5.916	irrational
$\sqrt{12.25}$	3.5	rational
$\sqrt{2.25}$	1.5	rational
$\sqrt{6}$	2.449	irrational
$\sqrt{16}$	4	rational
$\sqrt{26}$	5.099	irrational

7. a. $\sqrt{-36}$ has no real number solution

   b. $-\sqrt{81} = -9$

   c. $\pm\sqrt{144} = \pm12$

9. a. $\sqrt{49} = 7$

   b. $-\sqrt{225} = -15$

   c. $\pm\sqrt{400} = \pm20$

11.

x	$y = \sqrt{x+4}$
-4	$\sqrt{-4+4} = 0$
-2	$\sqrt{-2+4} = \sqrt{2}$
0	$\sqrt{0+4} = 2$
2	$\sqrt{2+4} = \sqrt{6}$
4	$\sqrt{4+4} = \sqrt{8}$

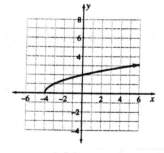

Section 8.2 (con't)

13.

x	$y = \sqrt{x-2}$
-4	$\sqrt{-4-2} = \sqrt{-6}$ not a real number
-2	$\sqrt{-2-2} = \sqrt{-4}$ not a real number
0	$\sqrt{0-2} = \sqrt{-2}$ not a real number
2	$\sqrt{2-2} = 0$
4	$\sqrt{4-2} = \sqrt{2}$

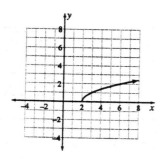

15.

x	$y = \sqrt{2x}$
-4	$\sqrt{2(-4)} = \sqrt{-8}$ not a real number
-2	$\sqrt{2(-2)} = \sqrt{-4}$ not a real number
0	$\sqrt{2(0)} = 0$
2	$\sqrt{2(2)} = 2$
4	$\sqrt{2(4)} = \sqrt{8}$

17. a. $\sqrt{5} \cdot \sqrt{3} = \sqrt{5 \cdot 3} = \sqrt{15}$

b. $\left(3\sqrt{5}\right)^2 = 3^2(\sqrt{5})^2 = 9 \cdot 5 = 45$

c. $\sqrt{16 \cdot 3} = \sqrt{16} \cdot \sqrt{3} = 4\sqrt{3}$

19. a. $\left(3\sqrt{7}\right)^2 = 3^2\left(\sqrt{7}\right)^2 = 9 \cdot 7 = 63$

b. $\sqrt{13}\sqrt{13} = (\sqrt{13})^2 = 13$

c. $\sqrt{36 \cdot 2} = \sqrt{36}\sqrt{2} = 6\sqrt{2}$

21. a. $\sqrt{3} \cdot \sqrt{27} = \sqrt{3 \cdot 27} = \sqrt{81} = 9$

b. $\left(2\sqrt{5}\right)^2 = 2^2(\sqrt{5})^2 = 4 \cdot 5 = 20$

c. $\sqrt{18} = \sqrt{9 \cdot 2} = \sqrt{9} \cdot \sqrt{2} = 3\sqrt{2}$

23. a. $\sqrt{a}\sqrt{a} = \left(\sqrt{a}\right)^2 = a$

b. $\sqrt{b^2} = b$

c. $\sqrt{121a^2} = \sqrt{121} \cdot \sqrt{a^2} = 11a$

25. a. $\sqrt{32a} \cdot \sqrt{2a} = \sqrt{32a \cdot 2a}$

$= \sqrt{64a^2} = 8a$

b. $\sqrt{2x} \cdot \sqrt{2x} = (\sqrt{2x})^2 = 2x$

c. $\sqrt{16b^2} = \sqrt{16}\sqrt{b^2} = 4b$

Section 8.2 (con't)

27. a. $\sqrt{\dfrac{x^2}{9}} = \dfrac{\sqrt{x^2}}{\sqrt{9}} = \dfrac{x}{3}$

b. $\sqrt{\dfrac{4}{25}} = \dfrac{\sqrt{4}}{\sqrt{25}} = \dfrac{2}{5}$

c. $\sqrt{\dfrac{45}{5}} = \sqrt{9} = 3$

29. a. $\sqrt{\dfrac{28x}{7x^3}} = \sqrt{\dfrac{4}{x^2}} = \dfrac{\sqrt{4}}{\sqrt{x^2}} = \dfrac{2}{x}$

b. $\sqrt{\dfrac{3x^2}{27}} = \sqrt{\dfrac{x^2}{9}} = \dfrac{\sqrt{x^2}}{\sqrt{9}} = \dfrac{x}{3}$

c. $\sqrt{\dfrac{8a^4}{32}} = \sqrt{\dfrac{a^4}{4}} = \dfrac{\sqrt{a^4}}{\sqrt{4}} = \dfrac{a^2}{2}$

31. a. $\sqrt{\dfrac{8x^2y}{2y}} = \sqrt{4x^2} = 2x$

b. $\sqrt{\dfrac{2x^4}{50y^2}} = \sqrt{\dfrac{x^4}{25y^2}} = \dfrac{\sqrt{x^4}}{\sqrt{25y^2}} = \dfrac{x^2}{5y}$

c. $-\sqrt{\dfrac{3xy^4}{27x}} = -\sqrt{\dfrac{y^4}{9}} = -\dfrac{\sqrt{y^4}}{\sqrt{9}} = -\dfrac{y^2}{3}$

33. a. $\{2.1, 2.9, 2\} = \{2, 2.1, 2.9\}$

$2^2 + 2.1^2 = 2.9^2$?

$4 + 4.41 = 8.41$, Pythagorean triple

b. $\{2, \sqrt{3}, 1\} = \{1, \sqrt{3}, 2\}$

$1^2 + (\sqrt{3})^2 = 2^2$?

$1 + 3 = 4$, Pythagorean triple

33. c. $1^2 + 2^2 = 3^2$?

$1 + 4 \neq 9$, not a Pythagorean triple

35. a. $\{3, \sqrt{7}, 4\} = \{\sqrt{7}, 3, 4\}$

$(\sqrt{7})^2 + 3^2 = 4^2$?

$7 + 9 = 16$, Pythagorean triple

b. $(-5)^2 + (-12)^2 = (-13)^2$?

$25 + 144 = 169$, Pythagorean triple but does not satisfy the Pythagorean theorem because it does not represent a triangle.

c. $\{6, \sqrt{13}, 7\} = \{\sqrt{13}, 6, 7\}$

$(\sqrt{13})^2 + 6^2 = 7^2$?

$13 + 36 = 49$, Pythagorean triple

37. a. $\sqrt{(4-2)^2 + (9-3)^2} = \sqrt{2^2 + 6^2}$

$= \sqrt{4 + 36} = \sqrt{40} = \sqrt{4(10)} = 2\sqrt{10}$

b. $\dfrac{9-3}{4-2} = \dfrac{6}{2} = 3$

c. $b = y - mx$, $b = 3 - 3(2)$, $b = 3 - 6$,

$b = -3$, $y = 3x - 3$

Section 8.2 (con't)

39. a. $\sqrt{(5-2)^2+(-1-2)^2}=\sqrt{3^2+(-3)^2}$

$=\sqrt{9+9}=\sqrt{18}=\sqrt{9(2)}=3\sqrt{2}$

b. $\dfrac{-1-2}{5-2}=\dfrac{-3}{3}=-1$

c. b = 2 - (-1)(2), b = 2 + 2, b = 4

y = -x + 4

41. a. $\sqrt{(4-(-3))^2+(2-3)^2}=\sqrt{7^2+(-1)^2}$

$=\sqrt{49+1}=\sqrt{50}=\sqrt{(2)(25)}=5\sqrt{2}$

b. $\dfrac{2-3}{4-(-3)}=\dfrac{-1}{7}=-\dfrac{1}{7}$

c. b = 3 - $(-\frac{1}{7})$(-3), b = $2\frac{4}{7}$

y = $-\frac{1}{7}$x + $2\frac{4}{7}$

43. a. $\sqrt{(3-(-3))^2+(-3-(-1))^2}=\sqrt{6^2+(-2)^2}$

$=\sqrt{36+4}=\sqrt{40}=\sqrt{4(10)}=2\sqrt{10}$

b. $\dfrac{-3-(-1)}{3-(-3)}=\dfrac{-2}{6}=-\dfrac{1}{3}$

c. b = -3 - $(-\frac{1}{3})$(3), b = -2

y = $-\frac{1}{3}$x - 2

45. Let A = (3, 4), B = (0, 1) and C = (6, 1)

AB = $\sqrt{(0-3)^2+(1-4)^2}$, AB = $\sqrt{9+9}$,

AB = $\sqrt{18}$

45. BC = $\sqrt{(6-0)^2+(1-1)^2}$, BC = $\sqrt{36+0}$,

BC = 6

AC = $\sqrt{(6-3)^2+(1-4)^2}$, AC = $\sqrt{9+9}$,

AC = $\sqrt{18}$,

$(\sqrt{18})^2+(\sqrt{18})^2=6^2$?
$18+18=36$

AB = AC and sides make a Pythagorean

triple, this is an isosceles right triangle.

47. Let A = (6, 5), B = (4, 2) and C = (8, 2)

AB = $\sqrt{(4-6)^2+(2-5)^2}$, AB = $\sqrt{4+9}$,

AB = $\sqrt{13}$

BC = $\sqrt{(8-4)^2+(2-2)^2}$, BC = $\sqrt{16+0}$

BC = 4

AC = $\sqrt{(8-6)^2+(2-5)^2}$, AC = $\sqrt{4+9}$,

AC = $\sqrt{13}$

$(\sqrt{13})^2+(\sqrt{13})^2=4^2$?, $13+13\neq16$

AB = AC, sides do not make a

Pythagorean triple, this is an isosceles

triangle.

Section 8.2 (con't)

49. Let A = (4, 8), B = (1, 6) and C = (5, 0)

$$AB = \sqrt{(1-4)^2 + (6-8)^2}, AB = \sqrt{9+4},$$

$$AB = \sqrt{13}$$

$$BC = \sqrt{(5-1)^2 + (0-6)^2},$$

$$BC = \sqrt{16+36}, \; BC = \sqrt{52},$$

$$AC = \sqrt{(5-4)^2 + (0-8)^2},$$

$$AC = \sqrt{1+64}, AC = \sqrt{65}$$

$$\left(\sqrt{13}\right)^2 + \left(\sqrt{52}\right)^2 = \left(\sqrt{65}\right)^2 ?$$

$$13 + 52 = 65$$

Sides make a Pythagorean triple, this is a right triangle.

51. a. $25^0 = 1$

b. $25^{-1} = \frac{1}{25}$

c. $25^{\frac{1}{2}} = \sqrt{25} = 5$

d. $25^{0.5} = \sqrt{25} = 5$

53. a. $9^{\frac{1}{2}} = \sqrt{9} = 3$

b. $9^0 = 1$

c. $9^{0..5} = \sqrt{9} = 3$

d. $9^{-1} = \frac{1}{9}$

55. a. $\left(\frac{1}{4}\right)^{-1} = 4$

b. $\left(\frac{1}{4}\right)^0 = 1$

c. $\left(\frac{1}{4}\right)^{\frac{1}{2}} = \sqrt{\frac{1}{4}} = \frac{\sqrt{1}}{\sqrt{4}} = \frac{1}{2}$

d. $\left(\frac{1}{4}\right)^{0.5} = \sqrt{\frac{1}{4}} = \frac{\sqrt{1}}{\sqrt{4}} = \frac{1}{2}$

57. a. $(0.25)^{-1} = \left(\frac{1}{4}\right)^{-1} = 4$

b. $(0.01)^{0.5} = \left(\frac{1}{100}\right)^{0.5} = \sqrt{\frac{1}{100}} = \frac{1}{10} = 0.1$

c. $(6.25)^{0.5} = \left(\frac{625}{100}\right)^{0.5} = \sqrt{\frac{625}{100}} = \frac{\sqrt{625}}{\sqrt{100}}$

$= \frac{25}{10} = 2.5$

d. $(0.25)^{0.5} = \left(\frac{1}{4}\right)^{0..5} = \sqrt{\frac{1}{4}} = \frac{1}{2} = 0.5$

e. $(0.02)^{-1} = \left(\frac{2}{100}\right)^{-1} = \frac{100}{2} = 50$

f. $(0.05)^{-1} = \left(\frac{5}{100}\right)^{-1} = \frac{100}{5} = 20$

59. $y = \sqrt{(-2)+3}$, $y = 1$, yes this expression is defined when x = -2.

61. Yes, if $x \leq 0$ then -2x will be positive and the expression will have real number solutions.

63. True only if x > 1. For example $\sqrt{9} = 3$ and 3 < 9, but $\sqrt{\frac{1}{9}} = \frac{1}{3}$ and $\frac{1}{3} > \frac{1}{9}$.

Section 8.2 (con't)

65. Let n = 0.5, then $\left(\dfrac{a}{b}\right)^n = \left(\dfrac{a}{b}\right)^{0.5}$

$\left(\dfrac{a}{b}\right)^{0.5} = \sqrt{\dfrac{a}{b}}$. and $\dfrac{a^{0.5}}{b^{0.5}} = \dfrac{\sqrt{a}}{\sqrt{b}}$,

since $\left(\dfrac{a}{b}\right)^{0.5} = \dfrac{a^{0.5}}{b^{0.5}}$, then $\sqrt{\dfrac{a}{b}} = \dfrac{\sqrt{a}}{\sqrt{b}}$.

Section 8.3

1. The radicand is the value under the radical sign.

   a. -3

   b. ab

   c. $4x^2$

   d. 2

   e. x

   f. $x+2$

3.

x	$f(x) = \sqrt{4-x}$
-4	$\sqrt{4-(-4)} = \sqrt{8} = 2\sqrt{2}$
-1	$\sqrt{4-(-1)} = \sqrt{5}$
0	$\sqrt{4-0} = 2$
4	$\sqrt{4-4} = 0$
6	$\sqrt{4-6} = \sqrt{-2}$, not a real #

5.

x	$f(x) = \sqrt{3-x}$
-4	$\sqrt{3-(-4)} = \sqrt{7}$
-1	$\sqrt{3-(-1)} = \sqrt{4} = 2$
0	$\sqrt{3-0} = \sqrt{3}$
4	$\sqrt{3-4} = \sqrt{-1}$, not a real #
6	$\sqrt{3-6} = \sqrt{-3}$, not a real #

7. a. $\sqrt{ab^2} = \sqrt{b^2}\sqrt{a} = |b|\sqrt{a}$

   b. $\sqrt{a^2 b} = \sqrt{a^2}\sqrt{b} = |a|\sqrt{b}$

   c. $\sqrt{a^2 b^2} = \sqrt{(ab)^2} = |ab|$

9. a. $\sqrt{49x^2} = \sqrt{49}\sqrt{x^2} = 7|x|$

   b. $\sqrt{121y^2} = \sqrt{121}\sqrt{y^2} = 11|y|$

11. a. $\sqrt{p^1 p^2} = \sqrt{p}\sqrt{p^2} = |p|\sqrt{p}$

    b. $\sqrt{p^4} = \sqrt{(p^2)^2} = p^2$

13. a. $\sqrt{b \cdot b^4} = \sqrt{b}\sqrt{(b^2)^2} = b^2\sqrt{b}$

    b. $\sqrt{c \cdot c^2} = \sqrt{c}\sqrt{c^2} = |c|\sqrt{c}$

15. While x could be any number, x^2 is always a positive number, since the principle square root is a positive number x must have an absolute value.

17 & 19   The expressions are defined when the radicands are ≥ 0.

17. a. x - 1 ≥ 0, x ≥ 1

    b. x + 3 ≥ 0, x ≥ -3

19. a. 4 - x ≥ 0, x ≤ 4

    b. 3 - x ≥ 0, x ≤ 3

21. The graphs intersect at x = 6.

Section 8.3 (con't)

23. Equation is defined when $x + 3 \geq 0$,

$x \geq -3$,

$\sqrt{x+3} = 1, \left(\sqrt{x+3}\right)^2 = 1^2,$
$x + 3 = 1, \; x = -2$

25. Equation is defined when $3 - x \geq 0$.

$x \leq 3$

$\sqrt{3-x} = 2, \left(\sqrt{3-x}\right)^2 = 2^2,$
$3 - x = 4, \; 3 = 4 + x, \; x = -1$

27. Defined when $x - 2 \geq 0, x \geq 2$

$\sqrt{x-2} = 3, \left(\sqrt{x-2}\right)^2 = 3^2,$
$x - 2 = 9, \; x = 11$

29. Defined when $2x + 2 \geq 0$,

$2x \geq -2, \; x \geq -1$

$\sqrt{2x+2} = 4, \left(\sqrt{2x+2}\right)^2 = 4^2,$
$2x + 2 = 16, \; 2x = 14, \; x = 7$

31. Defined when $3x - 3 \geq 0$,

$3x \geq 3, \; x \geq 1$

$\sqrt{3x-3} = 6, \left(\sqrt{3x-3}\right)^2 = 6^2,$
$3x - 3 = 36, \; 3x = 39, \; x = 13$

33. Defined when $5 - x \geq 0, x \leq 5$

$\sqrt{5-x} = 3, \left(\sqrt{5-x}\right)^2 = 3^2,$
$5 - x = 9, \; 5 = 9 + x, \; x = -4$

35 & 37. The formula for distance to the

horizon is $d \approx \sqrt{\dfrac{3h}{2}}$

35. $h = 1454$ ft., $d \approx \sqrt{\dfrac{3(1454)}{2}},$

$d \approx \sqrt{2181}, \; d \approx 46.7$ miles

37. $h = 1002$ ft., $d \approx \sqrt{\dfrac{3(1002)}{2}},$

$d \approx \sqrt{1503}, \; d \approx 38.8$ miles

39. $h = 24$ ft., $d \approx \sqrt{\dfrac{3(24)}{8}},$

$d \approx \sqrt{9}, \; d \approx 3$ miles

41. $h = 96$ ft., $d \approx \sqrt{\dfrac{3(96)}{8}},$

$d \approx \sqrt{36}, \; d \approx 6$ miles

43. a. $30 \approx \sqrt{\dfrac{3h}{2}}, 30^2 \approx \left(\sqrt{\dfrac{3h}{2}}\right)^2,$

$900 \approx \dfrac{3h}{2}, \; 1800 \approx 3h, \; h \approx 600$ ft.

b. $30 \approx \sqrt{\dfrac{3h}{8}}, 30^2 \approx \left(\sqrt{\dfrac{3h}{8}}\right)^2,$

$900 \approx \dfrac{3h}{8}, \; 7200 \approx 3h, \; h \approx 2400$ ft.

Section 8.3 (con't)

43. c. $2400 \div 600 = 4$ times higher

   d. The moon's radius is smaller.

45 to 51 Use $t = \sqrt{\dfrac{2d}{g}}$ for the time it takes

an object to fall a distance, d, and $v = gt$ for the speed of the object when it hits the ground. Remember $g = 32.2$ ft per $\sec^2$.

45. $d = 1821$ ft,

$$t = \sqrt{\dfrac{2(1821\,\text{ft})}{32.2\,\dfrac{\text{ft}}{\sec^2}}}, \; t \approx \sqrt{113.11\,\sec^2}$$

$t \approx 10.6$ sec.

$v \approx (32.2\ \text{ft/sec}^2)(\sqrt{113.11}\ \sec),$

$v \approx 342.5$ ft per sec

47. $d = 1063$ ft,

$$t = \sqrt{\dfrac{2(1063\,\text{ft})}{32.2\,\dfrac{\text{ft}}{\sec^2}}}, \; t \approx \sqrt{66\,\sec^2},$$

$t \approx 8.1$ sec.

$v \approx (32.2\ \text{ft/sec}^2)(\sqrt{66}\ \sec),$

$v \approx 261.6$ ft per sec

49. $12\ \sec = \sqrt{\dfrac{2d}{32.2\,\dfrac{\text{ft}}{\sec^2}}},$

$$12^2\ \sec^2 = \left(\sqrt{\dfrac{2d}{32.2\,\dfrac{\text{ft}}{\sec^2}}} \right)^2$$

$$144\ \sec^2 = \dfrac{2d}{32.2\,\dfrac{\text{ft}}{\sec^2}}, \; 4636.8\,\text{ft} = 2d,$$

$d = 2318.4$ ft

51. 1 mile $= 5280$ ft

$$t = \sqrt{\dfrac{2(5280\,\text{ft})}{32.2\,\dfrac{\text{ft}}{\sec^2}}}, \; t \approx \sqrt{327.95\ \sec}$$

$t \approx 18.1$ sec

53. $E = \sqrt{W \cdot R}, \; E^2 = \left(\sqrt{WR} \right)^2,$

$E^2 = WR, \; R = \dfrac{E^2}{W}$

55. $t = \sqrt{\dfrac{2d}{g}}, \; t^2 = \left(\sqrt{\dfrac{2d}{g}} \right)^2, \; t^2 = \dfrac{2d}{g},$

$gt^2 = 2d, \; g = \dfrac{2d}{t^2}$

57. $V_0 = \sqrt{\dfrac{GM}{R}}, \; (V_0)^2 = \left(\sqrt{\dfrac{GM}{R}} \right)^2,$

$V_0^2 = \dfrac{GM}{R}, \; RV_0^2 = GM, \; M = \dfrac{RV_0^2}{G}$

Section 8.3 (con't)

59. $\sqrt{\dfrac{3h_e}{2}} = \sqrt{\dfrac{3h_m}{8}}, \left(\sqrt{\dfrac{3h_e}{2}}\right)^2 = \left(\sqrt{\dfrac{3h_m}{8}}\right)^2$

$\dfrac{3h_e}{2} = \dfrac{3h_m}{8}, 24h_e = 6h_m, \dfrac{h_e}{h_m} = \dfrac{6}{24}$

$\dfrac{h_e}{h_m} = \dfrac{1}{4}$

61. If the expression could be less than zero use an absolute value.

Mid-Chapter 8 Test

1. **a.** $4^2 + 6^2 = 8^2$?

    $16 + 36 \neq 64$, not a right triangle

    b. $10^2 + 15^2 = 20^2$?

    $100 + 225 \neq 400$, not a right triangle

    c. $15^2 + 20^2 = 25^2$?

    $225 + 400 = 625$, right triangle

2. $\sqrt{120^2 + 50^2} = \sqrt{16900} = 130$ miles

3. $\sqrt{6^2 + 15^2} = \sqrt{261}$ inches

4. **a.** $\sqrt{2} \cdot \sqrt{18} = \sqrt{2 \cdot 18} = \sqrt{36} = 6$

    b. $\left(3\sqrt{2}\right)^2 = 3^2\left(\sqrt{2}\right)^2 = 9 \cdot 2 = 18$

    c. $\sqrt{72} = \sqrt{36 \cdot 2} = 6\sqrt{2}$

    d. $\left(2\sqrt{3}\right)^2 = 2^2\left(\sqrt{3}\right)^2 = 4 \cdot 3 = 12$

    e. $\sqrt{48} = \sqrt{16 \cdot 3} = 4\sqrt{3}$

    f. $-\sqrt{4} = -2$

    g. $\sqrt{-16}$, not a real number

    h. $\pm\sqrt{\dfrac{25}{16}} = \pm\dfrac{\sqrt{25}}{\sqrt{16}} = \pm\dfrac{5}{4}$

    i. $\dfrac{\sqrt{2}}{\sqrt{32}} = \sqrt{\dfrac{2}{32}} = \sqrt{\dfrac{1}{16}} = \dfrac{\sqrt{1}}{\sqrt{16}} = \dfrac{1}{4}$

5. **a.** $\sqrt{3x}\sqrt{27x^3} = \sqrt{81x^4} = 9x^2$

    b. $\sqrt{\dfrac{3x^2}{12y^4}} = \sqrt{\dfrac{x^2}{4y^4}} = \dfrac{\sqrt{x^2}}{\sqrt{4y^4}} = \dfrac{|x|}{2y^2}$

    c. $\sqrt{\dfrac{196x}{x^3}} = \sqrt{\dfrac{196}{x^2}} = \dfrac{\sqrt{196}}{\sqrt{x^2}} = \dfrac{14}{|x|}$

    d. $\sqrt{a^2 + b^2}$, is already simplified

6. $\sqrt{(-4-2)^2 + (6-(-3))^2} = \sqrt{36 + 81}$

    $\sqrt{117} = \sqrt{9 \cdot 13} = 3\sqrt{13} \approx 10.8$

7. $11^2 = 121$, $12^2 = 144$

    $\sqrt{135}$ is between 11 and 12

8. $I = \sqrt{\dfrac{W}{R}}$, $I^2 = \left(\sqrt{\dfrac{W}{R}}\right)^2$,

    $I^2 = \dfrac{W}{R}$, $R = \dfrac{W}{I^2}$

9. $\sqrt{x-3} = 2$, $\left(\sqrt{x-3}\right)^2 = 2^2$, $x - 3 = 4$,

    $x = 7$;

    Equation is defined when $x - 3 \geq 0$, $x \geq 3$.

10. There are no values for y when the radicand is less than zero.

Section 8.4

1. From the graph the x-intercept points are
   (-3, 0) and (2, 0), the y-intercept point is
   (0, -6).

   The axis of symmetry is half-way
   between -3 and 2, at $x = (-3 + 2) \div 2$,
   $x = -0.5$.

   When $x = -0.5$, $y = (-0.5)^2 + 0.5 - 6$,
   $y = -6.25$. The vertex is (-0.5, -6.25).

3. From the graph the x-intercept points are
   (0, 0) and (5, 0), the y-intercept point is
   (0, 0).

   The axis of symmetry is half-way
   between 0 and 5 at $x = (5 - 0) \div 2$,
   $x = 2.5$

   When $x = 2.5$, $y = 5(2.5) - (2.5)^2$,
   $y = 6.25$. The vertex is (2.5, 6.25).

5.

x	$y = x^2 - 6x + 7$
0	$0^2 - 6(0) + 7 = 7$
1	$1^2 - 6(1) + 7 = 2$
2	$2^2 - 6(2) + 7 = -1$
3	$3^2 - 6(3) + 7 = -2$
4	$4^2 - 6(4) + 7 = -1$
5	$5^2 - 6(5) + 7 = 2$

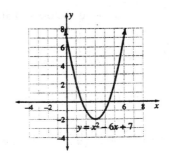

The x-intercepts are between 1 and 2,
and 4 and 5. Using a graphing calculator
we find the x-intercept points at
approximately (1.6, 0) and (4.4, 0). The
y-intercept point is (0, 7).

The axis of symmetry is $x = 3$.

When $x = 3$, $y = -2$, the vertex is (3, -2)

Section 8.4 (con't)

7.

x	$y = x^2 + x - 12$
-4	$(-4)^2 + (-4) - 12 = 0$
-3	$(-3)^2 + (-3) - 12 = -6$
-1	$(-1)^2 + (-1) - 12 = -12$
0	$0^2 + 0 - 12 = -12$
1	$1^2 + 1 - 12 = -10$
3	$3^2 + 3 - 12 = 0$

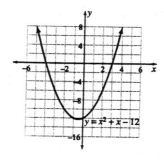

The x-intercept points are at (-4, 0) and (3, 0), the y-intercept point is at (0, -12).

The axis of symmetry is half-way between -1 and 0 at x = -0.5.

When x = -0.5, $y = (-0.5)^2 + (-0.5) - 12$, y = -12.25. The vertex is (-0.5, -12.25)

9.

x	$y = x - x^2$
-1	$(-1) - (-1)^2 = -2$
0	$0 - 0^2 = 0$
1	$1 - 1^2 = 0$
2	$2 - 2^2 = -2$

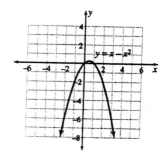

The x-intercept points are at (0, 0) and (1, 0), the y-intercept point is (0, 0).

The axis of symmetry is half-way between 0 and 1 at x = 0.5.

When x = 0.5, $y = 0.5 - (0.5)^2$, y = 0.25. The vertex is (0.5, 0.25).

Section 8.4 (con't)

11.

x	$y = 4x - 2x^2$
-1	$4(-1) - 2(-1)^2 = -6$
0	$4(0) - 2(0)^2 = 0$
1	$4(1) - 2(1)^2 = 2$
2	$4(2) - 2(2)^2 = 0$
3	$4(3) - 2(3)^2 = -6$

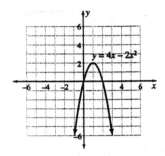

The x-intercept points are at (0, 0) and (2, 0), the y-intercept point is (0, 0).

The axis of symmetry is at x = 1.

When x = 1, y = 2. The vertex is (1, 2).

13. The range is all y values or $y \geq -6.25$.

15. The range is all y values or $y \leq 6.25$.

17. In $y = 2x^2 + 3x + 1$, a = 2, b = 3, c = 1

19. In $y = r^2 - 4r + 4$, a = 1, b = -4, c = 4

21. In $y = x^2 - 4$, a = 1, b = 0, c = -4

23. In $4t^2 - 8t = 0$, a = 4, b = -8, c = 0

25. In $4 = x - x^2$, add x^2 and subtract x from both sides to get $x^2 - x + 4 = 0$, a = 1, b = -1, c = 4

27. In $x^2 = x - 1$, subtract x and add 1 to both sides to get $x^2 - x + 1$, a = 1, b = -1, c = 1

29. In $h = -0.5gt^2 + vt + s$, a = -0.5g, b = v, c = s

31. In $A = \pi r^2$, a = π, b = 0, c = 0

33. When a = 4, b = 4 and c = 1, then

$y = 4x^2 + 4x + 1$

35. When a = 9, b = 0 and c = -16, then

$y = 9x^2 + 0x + (-16)$, $y = 9x^2 - 16$

37. When a = 3, b = 6 and c = 0, then

$y = 3x^2 + 6x + 0$, $y = 3x^2 + 6x$

39. a. When y = 6, x = {-4, 3}

b. When y = -4, x = {-2, 1}

c. When y = -8 there are no solutions.

d. When y = 0, x = {-3, 2}

41. a. When y = 6, x = {2, 3}

b. When y = 8 there are no solutions

c. When y = -6, x = {-1, 6}

d. When y = 0, x = {0, 5}

Section 8.4 (con't)

43. Solve $h = h_0 - \frac{1}{2}gt^2$, for t, $h + \frac{1}{2}gt^2 = h_0$,

$$\frac{1}{2}gt^2 = h_0 - h, \ t^2 = \frac{2(h_0 - h)}{g},$$

$$t = \sqrt{\frac{2(h_0 - h)}{g}}$$

Half-way is $1612 \div 2 = 806$ ft

$$t = \sqrt{\frac{2(1612 - 806)}{32.2}}, \ t \approx 7.1 \sec$$

At the bottom $h = 0$

$$t = \sqrt{\frac{2(1612 - 0)}{32.2}}, \ t \approx 10 \sec$$

Section 8.5

1. $x^2 = 5, \sqrt{x^2} = \sqrt{5}, x = \pm\sqrt{5}$

3. $2x^2 = 14, x^2 = 7, \sqrt{x^2} = \sqrt{7}, x = \pm\sqrt{7}$

5. $x^2 = \dfrac{4}{25}, \sqrt{x^2} = \sqrt{\dfrac{4}{25}}, x = \pm\dfrac{2}{5}$

7. $100x^2 = 4, x^2 = \dfrac{4}{100}, x^2 = \dfrac{1}{25},$

$\sqrt{x^2} = \sqrt{\dfrac{1}{25}}, x = \pm\dfrac{1}{5}$

9. $49x^2 - 225 = 0, 49x^2 = 225, x^2 = \dfrac{225}{49},$

$\sqrt{x^2} = \sqrt{\dfrac{225}{49}}, x = \pm\dfrac{15}{7}, x = \pm2\tfrac{1}{7}$

11. $36x^2 = 121, x^2 = \dfrac{121}{36}, \sqrt{x^2} = \sqrt{\dfrac{121}{36}}$

$x = \pm\dfrac{11}{6}, x = \pm1\tfrac{5}{6}$

13. $75x^2 - 27 = 0, 75x^2 = 27, x^2 = \dfrac{27}{75},$

$x^2 = \dfrac{9}{25}, \sqrt{x^2} = \sqrt{\dfrac{9}{25}}, x = \pm\dfrac{3}{5}$

15. $\dfrac{3}{x} = \dfrac{x}{27}, x^2 = 81, \sqrt{x^2} = \sqrt{81}, x = \pm9$

17. $\dfrac{x}{4} = \dfrac{9}{x}, x^2 = 36, \sqrt{x^2} = \sqrt{36}, x = \pm6$

19. $(x - 4)(x + 4) = 0,$

$x - 4 = 0, \ x = 4; x + 4 = 0, x = -4$

21. $(2x - 1)(3x + 2) = 0,$

$2x - 1 = 0, 2x = 1, x = \tfrac{1}{2}$

$3x + 2 = 0, 3x = -2, x = -\tfrac{2}{3}$

23. $(2x - 5)(x + 2) = 0,$

$2x - 5 = 0, 2x = 5, x = \tfrac{5}{2}$

$x + 2 = 0, x = -2$

25. $x^2 - 4 = 0, \ (x + 2)(x - 2) = 0;$

$x + 2 = 0, x = -2$

$x - 2 = 0, x = 2$

27. $x^2 + x - 6 = 0, (x - 2)(x + 3) = 0;$

$x - 2 = 0, x = 2$

$x + 3 = 0, x = -3$

29. $x^2 - 2x - 15 = 0, (x - 5)(x + 3) = 0;$

$x - 5 = 0, x = 5$

$x + 3 = 0, x = -3$

31. $x^2 + 3x = 0, x(x + 3) = 0;$

$x = 0$

$x + 3 = 0, x = -3$

33. $2x^2 = -x, 2x^2 + x = 0, x(2x + 1) = 0;$

$x = 0$

$2x + 1 = 0, \ 2x = -1, \ x = -\tfrac{1}{2}$

Section 8.5 (con't)

35. $x^2 - 4x = 12$, $x^2 - 4x - 12 = 0$,

$(x - 6)(x + 2) = 0$;

$x - 6 = 0$, $x = 6$

$x + 2 = 0$, $x = -2$

37. $2x^2 = x + 3$, $2x^2 - x - 3 = 0$,

$(2x - 3)(x + 1) = 0$;

$2x - 3 = 0$, $2x = 3$, $x = \frac{3}{2}$

$x + 1 = 0$, $x = -1$

39. $x^2 = x + 12$, $x^2 - x - 12 = 0$,

$(x - 4)(x + 3) = 0$;

$x - 4 = 0$, $x = 4$

$x + 3 = 0$, $x = -3$

41. $x^2 + 6 = 7x$, $x^2 - 7x + 6 = 0$,

$(x - 6)(x - 1) = 0$;

$x - 6 = 0$, $x = 6$

$x - 1 = 0$, $x = 1$

43. $2x^2 + 3x = 5$, $2x^2 + 3x - 5 = 0$,

$(2x + 5)(x - 1) = 0$;

$2x + 5 = 0$, $2x = -5$, $x = -\frac{5}{2}$

$x - 1 = 0$, $x = 1$

45. $4x^2 - 25 = 0$, $(2x + 5)(2x - 5) = 0$;

$2x + 5 = 0$, $2x = -5$, $x = -\frac{5}{2}$

$2x - 5 = 0$, $2x = 5$, $x = \frac{5}{2}$

47. $3x^2 - 12 = 0$, $3(x^2 - 4) = 0$,

$x^2 - 4 = 0$, $(x + 2)(x - 2) = 0$;

$x + 2 = 0$, $x = -2$

$x - 2 = 0$, $x = 2$

49. $\frac{x + 2}{2} = \frac{5}{x - 1}$, $(x + 2)(x - 1) = 10$,

$x^2 + x - 2 = 10$, $x^2 + x - 12 = 0$,

$(x + 4)(x - 3) = 0$;

$x + 4 = 0$, $x = -4$

$x - 3 = 0$, $x = 3$

51. $\frac{1}{x - 3} = \frac{x - 2}{2}$, $(x - 3)(x - 2) = 2$,

$x^2 - 5x + 6 = 2$, $x^2 - 5x + 4 = 0$,

$(x - 1)(x - 4) = 0$;

$x - 1 = 0$, $x = 1$

$x - 4 = 0$, $x = 4$

53. $-16t^2 + 48t = 0$, $-16(t^2 - 3t) = 0$,

$t^2 - 3t = 0$, $t(t - 3) = 0$;

$t = 0$ sec

$t - 3 = 0$, $t = 3$ sec

Answers agree with graph

Section 8.5 (con't)

55. From the graph when y = 48 ft,

   $t = \{1, 3\}$

   $-16t^2 + 64t = 48$, $-16t^2 + 64t - 48 = 0$,

   dividing both sides by -16:

   $t^2 - 4t + 3 = 0$, $(t - 3)(t - 1) = 0$;

   $t - 3 = 0$, $t = 3$

   $t - 1 = 0$, $t = 1$

   Results agree

57. $-16t^2 + 64t = 0$, $t^2 - 4t = 0$, $t(t - 4) = 0$;

   $t = 0$

   $t - 4 = 0$, $t = 4$

   Results agree with graph.

59. $\sqrt{4 - x} = x - 2$, $\left(\sqrt{4 - x}\right)^2 = (x - 2)^2$,

   $4 - x = x^2 - 4x + 4$, $0 = x^2 - 3x$,

   $x(x - 3) = 0$;

   $x = 0$

   $x - 3 = 0$, $x = 3$

   $\sqrt{4 - 0} = 0 - 2$, false
   $\sqrt{4 - 3} = 3 - 2$, true

   discard $x = 0$,

   $x = 3$ is the solution

61. $\sqrt{x + 5} = x + 3$, $\left(\sqrt{x + 5}\right)^2 = (x + 3)^2$

   $x + 5 = x^2 + 6x + 9$, $0 = x^2 + 5x + 4$,

   $(x + 4)(x + 1) = 0$;

   $x + 4 = 0$, $x = -4$

   $x + 1 = 0$, $x = -1$

   $\sqrt{-4 + 5} = -4 + 3$, false
   $\sqrt{-1 + 5} = -1 + 3$, true

   discard $x = -4$,

   $x = -1$ is the solution

63. **a.** $A = (0, 0)$, $B = (4, 16)$

   b. $x^2 = 4x$ is solved by inputs, x, at A & B.

   c. $x^2 = 4x$, $x^2 - 4x = 0$, $x(x - 4) = 0$;

   $x = 0$

   $x - 4 = 0$, $x = 4$

Section 8.5 (con't)

65.

a	b	$a^2 + b^2$	$(a+b)^2$	$\sqrt{(a+b)}$	$\sqrt{a} + \sqrt{b}$
4	9	$4^2 + 9^2$ $= 16 + 81 = 97$	$(4+9)^2$ $= 13^2 = 169$	$\sqrt{4+9} = \sqrt{13}$ ≈ 3.61	$\sqrt{4} + \sqrt{9}$ $= 2 + 3 = 5$
1	3	$1^2 + 3^3$ $= 1 + 9 = 10$	$(1+3)^2$ $= 4^2 = 16$	$\sqrt{1+3} = \sqrt{4} = 2$	$\sqrt{1} + \sqrt{3}$ $= 1 + \sqrt{3} \approx 2.73$
4	5	$4^2 + 5^2$ $= 16 + 25 = 41$	$(4+5)^2$ $= 9^2 = 81$	$\sqrt{4+5} = \sqrt{9} = 3$	$\sqrt{4} + \sqrt{5}$ $= 2 + \sqrt{5} \approx 4.24$
3	6	$3^2 + 6^2$ $= 9 + 36 = 45$	$(3+6)^2$ $= 9^2 = 81$	$\sqrt{3+6} = \sqrt{9} = 3$	$\sqrt{3} + \sqrt{6}$ ≈ 4.18

Section 8.6

1. In $9x^2 + 6x + 1 = 0$, $a = 9$, $b = 6$, $c = 1$

3. In $3x^2 - 9x = 0$, $a = 3$, $b = -9$, $c = 0$

5. Add $-4x$ and 3 to both sides of

$x^2 = 4x - 3$ to get $x^2 - 4x + 3 = 0$, $a = 1$,

$b = -4$, $c = 3$

7. Add 9 to both sides of $x^2 = -9$ to get

$x^2 + 9 = 0$, $a = 1$, $b = 0$, $c = 9$

9. $\dfrac{-(-4) - \sqrt{(-4)^2 - 4(1)(-12)}}{2 \cdot 1}$

$= \dfrac{4 - \sqrt{16 + 48}}{2} = \dfrac{4 - \sqrt{64}}{2} = \dfrac{4 - 8}{2} = -2$

11. $\dfrac{-(-5) + \sqrt{(-5)^2 - 4(6)(1)}}{2(6)}$

$= \dfrac{5 + \sqrt{25 - 24}}{12} = \dfrac{5 + \sqrt{1}}{12} = \dfrac{6}{12} = \dfrac{1}{2}$

13. $\dfrac{-1 - \sqrt{(1)^2 - 4(3)(-4)}}{2(3)}$

$= \dfrac{-1 - \sqrt{1 + 48}}{6} = \dfrac{-1 - \sqrt{49}}{6} = \dfrac{-8}{6} = -\dfrac{4}{3}$

15. $\dfrac{-6 + \sqrt{(6)^2 - 4(16)(-1)}}{2(16)}$

$= \dfrac{-6 + \sqrt{36 + 64}}{32} = \dfrac{-6 + \sqrt{100}}{32} = \dfrac{4}{32} = \dfrac{1}{8}$

17. a. $\sqrt{2} \approx 1$, $\dfrac{2 - 1}{2} = \dfrac{1}{2}$, $\dfrac{2 - \sqrt{2}}{2} \approx 0.293$

17. b. $\sqrt{6} \approx 2$, $\dfrac{3 + 2}{3} = \dfrac{5}{3}$, $\dfrac{3 + \sqrt{6}}{3} \approx 1.816$

c. $\sqrt{6} \approx 2$, $\dfrac{3(2)}{3} = 2$, $\dfrac{3\sqrt{6}}{3} = \sqrt{6} \approx 2.449$

19. a. $\sqrt{5} \approx 2$, $\dfrac{3 - 2}{3} = \dfrac{1}{3}$, $\dfrac{3 - \sqrt{5}}{3} \approx 0.255$

b. $\sqrt{10} \approx 3$, $\dfrac{2(3)}{2} = 3$, $\dfrac{2\sqrt{10}}{2} \approx 3.162$

c. $\sqrt{10} \approx 3$, $\dfrac{2 + 3}{2} = \dfrac{5}{2}$, $\dfrac{2 + \sqrt{10}}{2} \approx 2.581$

21. a. $\sqrt{13} \approx 4$, $\dfrac{-5 - 4}{2} = \dfrac{-9}{2}$,

$\dfrac{-5 - \sqrt{13}}{2} \approx -4.303$

b. $\sqrt{57} \approx 8$, $\dfrac{-9 + 8}{4} = \dfrac{-1}{4}$,

$\dfrac{-9 + \sqrt{57}}{4} \approx -0.363$

23. a. $\sqrt{28} \approx 5$, $\dfrac{-2 - 5}{6} = \dfrac{-7}{6}$,

$\dfrac{-2 - \sqrt{28}}{6} \approx -1.215$

b. $\sqrt{84} \approx 9$, $\dfrac{-2 + 9}{10} = \dfrac{7}{10}$,

$\dfrac{-2 + \sqrt{84}}{10} \approx 0.717$

Section 8.6 (con't)

25. In $4x^2 + 3x - 1 = 0$, $a = 4$, $b = 3$, $c = -1$

$$x = \frac{-3 \pm \sqrt{3^2 - 4(4)(-1)}}{2(4)},$$

$$x = \frac{-3 \pm \sqrt{9 + 16}}{8}, \quad x = \frac{-3 \pm 5}{8},$$

$$x = \frac{-3 + 5}{8}, \quad x = \frac{2}{8}, \quad x = \frac{1}{4}$$

$$x = \frac{-3 - 5}{8}, \quad x = \frac{-8}{8}, \quad x = -1$$

27. Add -5x and -2 to both sides of

$7x^2 = 5x + 2$ to get $7x^2 - 5x - 2 = 0$;

$a = 7$, $b = -5$, $c = -2$

$$x = \frac{-(-5) \pm \sqrt{(-5)^2 - 4(7)(-2)}}{2(7)},$$

$$x = \frac{5 \pm \sqrt{25 + 56}}{14}, \quad x = \frac{5 \pm \sqrt{81}}{14}, \quad x = \frac{5 \pm 9}{14}$$

$$x = \frac{5 + 9}{14}, \quad x = \frac{14}{14}, \quad x = 1$$

$$x = \frac{5 - 9}{14}, \quad x = \frac{-4}{14}, \quad x = \frac{-2}{7}$$

29. Add $10x^2$ and -2 to both sides of

$x = 2 - 10x^2$ to get $10x^2 + x - 2 = 0$,

$a = 10$, $b = 1$, $c = -2$

$$x = \frac{-1 \pm \sqrt{1^2 - 4(10)(-2)}}{2(10)}, \quad x = \frac{-1 \pm \sqrt{1 + 80}}{20},$$

$$x = \frac{-1 \pm \sqrt{81}}{20}, \quad x = \frac{-1 \pm 9}{20}$$

$$x = \frac{-1 + 9}{20}, \quad x = \frac{8}{20}, \quad x = \frac{2}{5}$$

$$x = \frac{-1 - 9}{20}, \quad x = \frac{-10}{20}, \quad x = -\frac{1}{2}$$

31. $a = -16$, $b = 30$, $c = 150$

$$t = \frac{-30 \pm \sqrt{30^2 - 4(-16)(150)}}{2(-16)},$$

$$t = \frac{-30 \pm \sqrt{10500}}{-32},$$

$$t = \frac{-30 + \sqrt{10500}}{-32}, \quad t \approx -2.26$$

$$t = \frac{-30 - \sqrt{10500}}{-32}, \quad t \approx 4.14$$

Discard negative answer, negative time
is not acceptable.

Section 8.6 (con't)

33. $a = -16$, $b = 15$, $c = 150$

$$t = \frac{-15 \pm \sqrt{15^2 - 4(-16)(150)}}{2(-16)},$$

$$t = \frac{-15 \pm \sqrt{9825}}{-32}$$

$$t = \frac{-15 + \sqrt{9825}}{-32}, t \approx -2.63$$

$$t = \frac{-15 - \sqrt{9825}}{-32}, t \approx 3.57$$

Discard negative answer.

35. Subtract 200 from both sides to get:

$-16t^2 + 30t - 50 = 0$, $a = -16$, $b = 30$,

$c = -50$

$$t = \frac{-30 \pm \sqrt{30^2 - 4(-16)(-50)}}{2(-16)},$$

$$t = \frac{-30 \pm \sqrt{-2300}}{-32}$$

No real number solutions, negative radicand.

37. Add 40 to both sides to get:

$-16t^2 + 20t + 90$, $a = -16$, $b = 20$, $c = 90$

$$t = \frac{-20 \pm \sqrt{20^2 - 4(-16)(90)}}{2(-16)},$$

$$t = \frac{-20 \pm \sqrt{6160}}{-32},$$

$$t = \frac{-20 + \sqrt{6160}}{-32}, t \approx -1.83$$

$$t = \frac{-20 - \sqrt{6160}}{-32}, t \approx 3.08$$

Discard negative answer.

39. $a = 1$, $b = 1$, $c = 1$,

$$x = \frac{-1 \pm \sqrt{1^2 - 4(1)(1)}}{2(1)}, x = \frac{-1 \pm \sqrt{-3}}{2}$$

No real number solutions, negative radicand

41. $2x^2 + x - 1 = 0$, $(2x - 1)(x + 1) = 0$,

$2x - 1 = 0$, $2x = 1$, $x = \frac{1}{2}$

$x + 1 = 0$, $x = -1$

43. $3x^2 + x - 2 = 0$, $(3x - 2)(x + 1) = 0$,

$3x - 2 = 0$, $3x = 2$, $x = \frac{2}{3}$

$x + 1 = 0$, $x = -1$

Section 8.6 (con't)

45. $a = 1$, $b = 2$, $c = 2$

$$x = \frac{-2 \pm \sqrt{2^2 - 4(1)(2)}}{2(1)}, \; x = \frac{-2 \pm \sqrt{-4}}{2}$$

No real number solution, negative radicand.

47. $4x^2 = 10$, $x^2 = \frac{10}{4}$, $\sqrt{x^2} = \sqrt{\frac{10}{4}}$,

$$x = \pm\frac{\sqrt{10}}{2}$$

49. $4x^2 - 9 = 0$, $(2x + 3)(2x - 3) = 0$,

$2x + 3 = 0$, $2x = -3$, $x = \frac{-3}{2}$

$2x - 3 = 0$, $2x = 3$, $x = \frac{3}{2}$

51. $3x^2 + 2x = 6$, $3x^2 + 2x - 6 = 0$

$a = 3$, $b = 2$, $c = -6$

$$x = \frac{-2 \pm \sqrt{2^2 - 4(3)(-6)}}{2(3)},$$

$$x = \frac{-2 \pm \sqrt{76}}{6}, \; x = \frac{-2 + \sqrt{76}}{6}, \; x \approx 1.120$$

$$x = \frac{-2 - \sqrt{76}}{6}, \; x \approx -1.786$$

53. $a = 5$, $b = 4$, $c = 1$

$$x = \frac{-4 \pm \sqrt{4^2 - 4(5)(1)}}{2(5)}, \; x = \frac{-4 \pm \sqrt{-4}}{10}$$

No real number solution, negative radicand.

55. $5x^2 = 4 - 2x$, $5x^2 + 2x - 4 = 0$

$a = 5$, $b = 2$, $c = -4$

$$x = \frac{-2 \pm \sqrt{2^2 - 4(5)(-4)}}{2(5)},$$

$$x = \frac{-2 \pm \sqrt{84}}{10}, \; x = \frac{-2 + \sqrt{84}}{10}, \; x \approx 0.717$$

$$x = \frac{-2 - \sqrt{84}}{10}, \; x \approx -1.117$$

57. $x^2 + 5 = -2x$, $x^2 + 2x + 5 = 0$

$a = 1$, $b = 2$, $c = 5$

$$x = \frac{-2 \pm \sqrt{2^2 - 4(1)(5)}}{2(1)}, \; x = \frac{-2 \pm \sqrt{-16}}{2}$$

No real number solution, negative radicand.

59. $x^2 + 9 = 6x$, $x^2 - 6x + 9 = 0$,

$(x - 3)^2 = 0$, $x - 3 = 0$, $x = 3$

61. $4x^2 = 4x + 7$, $4x^2 - 4x - 7 = 0$,

$a = 4$, $b = -4$, $c = -7$

$$x = \frac{-(-4) \pm \sqrt{(-4)^2 - 4(4)(-7)}}{2(4)},$$

$$x = \frac{4 \pm \sqrt{128}}{8},$$

$$x = \frac{4 + \sqrt{128}}{8}, \; x \approx 1.914$$

$$x = \frac{4 - \sqrt{128}}{8}, \; x \approx -0.914$$

Section 8.6 (con't)

63. $4x^2 = 12x - 9,\ 4x^2 - 12x + 9 = 0,$

$(2x - 3)^2 = 0,$

$2x - 3 = 0,\ 2x = 3,\ x = 1.5$

65. The number of solutions is the number of x-intercepts.

67. Answers will vary.

69. Let x = one number, the next consecutive number = x + 1

$y = x(x + 1),$

$x(x + 1) = 0,\ x = 0$

$x + 1 = 0,\ x = -1$

The graph opens up so the output is negative between the x-intercepts on the interval (-1, 0).

Section 8.7

1. The sum of the set is $156,000 and there are 6 items. Mean = $\dfrac{\$156,000}{6}$,

Mean = $26,000

Median is half-way between $12,000 and $13,000 at $12,500.

3. The sum of the set is $130,000 and there are 5 items. Mean = $\dfrac{\$130,000}{5}$,

Mean = $26,000

Median is $27,500

5. $Q_1 = \$10,000$, $Q_3 = \$13,000$

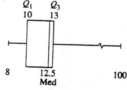

Numbers in thousands

7. $Q_1 = \dfrac{\$20,000 + \$27,500}{2}$, $Q_1 = \$23,750$

$Q_3 = \quad \$27,500$

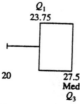

Numbers in thousands

9. Simplifying the equation:

$$s_x = \sqrt{\dfrac{324 + 256 + 196 + 169 + 169 + 5476}{5}}$$

$$s_x = \sqrt{\dfrac{6590}{5}}, \; s_x = \sqrt{1318}, \; s_x \approx 36.3 \text{ thousand}$$

$s \approx \$36,300$

11. Calculator gives; mean = $26,000,

$\sigma_x = \$3000$, $s_x = \$3354$

13. A few high-priced sales would raise the average.

15. No, not enough information is given to calculate the standard deviation.

17. One possible reason would be student's varying experience with exact measurements.

19. Answers will vary.

21. 2 standard deviations is 2(15) = 30

Normal readings will be from

3500°F - 30°F = 3470°F to

3500°F + 30°F = 3530°F.

23. 2 standard deviations is 2(0.013) = 0.026

Normal readings will be from

6.075 cm - 0.026 cm = 6.049 cm to

6.075 cm + 0.026 cm = 6.101 cm.

Section 8.7 (con't)

25. 2 standard deviations is 2(0.75) = 1.5

Normal wait will be from

2.5 min. - 1.5 min. = 1 min. to

2.5 min. + 1.5 min. = 4 min.

27. 1.5 standard deviations is

1.5(0.06) = 0.09

Normal outputs will be from

0.3 ppm - 0.09 ppm = 0.21 ppm to

0.3 ppm + 0.09 ppm = 0.39 ppm.

29. 0.5 standard deviations is 0.5(0.7) = 0.35

Normal flow will be from

$$48.4 \ \frac{gal}{min} - 0.35 \ \frac{gal}{min} = 48.05 \ \frac{gal}{min} \ \text{to}$$

$$48.4 \ \frac{gal}{min} + 0.35 \ \frac{gal}{min} = 48.75 \ \frac{gal}{min}.$$

Chapter 8 Review

1. a. $2^2 + 3^2 = 4^2$?,

$4 + 9 \neq 16$, not a right triangle

b. $5^2 + 17^2 = 18^2$?,

$25 + 289 \neq 324$, not a right triangle

c. $8^2 + 15^2 = 17^2$?,

$64 + 225 = 289$, right triangle

3. $\sqrt{6^2 + 24^2} = \sqrt{612} \approx 24.7$ ft.

5. a. $\sqrt{60} = \sqrt{4 \cdot 15} = 2\sqrt{15}$

b. $\sqrt{63} = \sqrt{9 \cdot 7} = 3\sqrt{7}$

c. $\sqrt{54} = \sqrt{9 \cdot 6} = 3\sqrt{6}$

7. a. $144^{\frac{1}{2}} = \sqrt{144} = 12$

b. $144^{-1} = \dfrac{1}{144}$

c. $144^0 = 1$

d. $144^{0.5} = \sqrt{144} = 12$

9. a. $(0.36)^{\frac{1}{2}} = \sqrt{0.36} = 0.6$

b. $(0.36)^{-1} = \left(\dfrac{36}{100}\right)^{-1} = \dfrac{100}{36} = \dfrac{25}{9}$

c. $(0.36)^{0.5} = \sqrt{0.36} = 0.6$

d. $(0.36)^0 = 1$

11. a. Sides:

$\sqrt{(-2-0)^2 + (3-0)^2} = \sqrt{13}$

$\sqrt{(-2-1)^2 + (3-5)^2} = \sqrt{13}$

$\sqrt{(1-3)^2 + (5-2)^2} = \sqrt{13}$

$\sqrt{(3-0)^2 + (2-0)^2} = \sqrt{13}$

Diagonals:

$\sqrt{(1-0)^2 + (5-0)^2} = \sqrt{26}$

$\sqrt{(-2-3)^2 + (3-2)^2} = \sqrt{26}$

Shape is a square.

b. Sides:

$\sqrt{(-4-0)^2 + (2-0)^2} = \sqrt{20} = 2\sqrt{5}$

$\sqrt{(-6-(-4))^2 + (-2-2)^2} = \sqrt{20} = 2\sqrt{5}$

$\sqrt{(-2-(-6))^2 + (-4-(-2))^2} = \sqrt{20} = 2\sqrt{5}$

$\sqrt{(-2-0)^2 + (-4-0)^2} = \sqrt{20} = 2\sqrt{5}$

Diagonals:

$\sqrt{(-6-0)^2 + (-2-0)^2} = \sqrt{40} = 2\sqrt{10}$

$\sqrt{(-2-(-4))^2 + (-4-2)^2} = \sqrt{40} = 2\sqrt{10}$

Shape is a square.

Chapter 8 Review (con't)

11. c. Sides:

$$\sqrt{(-1-0)^2+(3-0)^2}=\sqrt{10}$$

$$\sqrt{(-1-3)^2+(3-4)^2}=\sqrt{17}$$

$$\sqrt{(3-4)^2+(4-1)^2}=\sqrt{10}$$

$$\sqrt{(4-0)^2+(1-0)^2}=\sqrt{17}$$

Diagonals:

$$\sqrt{(3-0)^2+(4-0)^2}=\sqrt{25}=5$$

$$\sqrt{(-1-4)^2+(3-1)^2}=\sqrt{29}=3\sqrt{3}$$

13. a. $\sqrt{25x^2y^4}=\sqrt{25}\sqrt{x^2}\sqrt{(y^2)^2}=5xy^2$

b. $\sqrt{169x^6y^2}=\sqrt{169}\sqrt{(x^3)^2}\sqrt{y^2}$

$=13x^3y$

c. $\sqrt{2.25a^3}=\sqrt{2.25}\sqrt{a^2a}=1.5a\sqrt{a}$

d. $\sqrt{0.64b^5}=\sqrt{0.64}\sqrt{(b^2)^2b}=0.8b^2\sqrt{b}$

e. $\sqrt{\dfrac{80x^3}{5x}}=\sqrt{16x^2}=\sqrt{16}\sqrt{x^2}=4x$

f. $\sqrt{\dfrac{3a^4}{27b^6}}=\sqrt{\dfrac{a^4}{9b^6}}=\dfrac{\sqrt{(a^2)^2}}{\sqrt{9}\sqrt{(b^3)^2}}=\dfrac{a^2}{3b^3}$

g. $\sqrt{\dfrac{192a^6}{3}}=\sqrt{64a^6}=\sqrt{64}\sqrt{(a^3)^2}=8a^3$

h. $\sqrt{\dfrac{121}{49b^4}}=\dfrac{\sqrt{121}}{\sqrt{49}\sqrt{(b^2)^2}}=\dfrac{11}{7b^2}$

15.

n	$\left(\sqrt{2}\right)^n$
1	$\left(\sqrt{2}\right)^1=\sqrt{2}\approx1.41$
2	$\left(\sqrt{2}\right)^2=2$
3	$\left(\sqrt{2}\right)^3=2\sqrt{2}\approx2.83$
4	$\left(\sqrt{2}\right)^4=4$
5	$\left(\sqrt{2}\right)^5=4\sqrt{2}\approx5.66$
6	$\left(\sqrt{2}\right)^6=8$
7	$\left(\sqrt{2}\right)^7=8\sqrt{2}\approx11.3$
8	$\left(\sqrt{2}\right)^8=16$

17. a $\left(\sqrt{7x-3}\right)^2=5^2,\ 7x-3=25,$

$7x=28,\ x=4$

Defined when $7x-3\geq0,\ 7x\geq3,\ x\geq\frac{3}{7}$

b. $\left(\sqrt{2-x}\right)^2=(x-2)^2,$

$2-x=x^2-4x+4,\ x^2-3x+2=0,$
$(x-1)(x-2)=0,\ x-1=0,\ x=1$
$x-2=0,\ x=2$

$\sqrt{2-1}=1-2,$ false, discard solution
$\sqrt{2-2}=2-2,$ true

Solution is x = 2

Defined when 2 - x ≥ 0, x ≤ 2

Chapter 8 Review (con't)

17 c. $\left(\sqrt{4x-3}\right)^2 = 7^2$, $4x - 3 = 49$,

$4x = 52$, $x = 13$.

Defined when $4x - 3 \geq 0$, $4x \geq 3$,

$x \geq \frac{3}{4}$

19. a. $d \approx \sqrt{\dfrac{3(20)}{8}}$, $d \approx 2.7$ miles

b. $4 \approx \sqrt{\dfrac{3h}{8}}$, $4^2 \approx \left(\sqrt{\dfrac{3h}{8}}\right)^2$

$16 \approx \dfrac{3h}{8}$, $128 \approx 3h$, $h \approx \dfrac{128}{3}$, $h \approx 42\frac{2}{3}$

21. From the graph $x = 5$.

$\left(\sqrt{x-1}\right)^2 = 2^2$, $x - 1 = 4$, $x = 5$

23. Solutions are the x-intercepts. From the graph x = -2 and x = 4.

25. $x^2 = \frac{16}{144}$, $\sqrt{x^2} = \sqrt{\frac{16}{144}}$, $x = \pm\sqrt{\frac{1}{9}}$, $x = \pm\frac{1}{3}$

27. $x^2 - 5x + 4 = 0$, $(x - 4)(x - 1) = 0$,

x - 4 = 0, x = 4

x - 1 = 0, x = 1

29. a = 2, b = 5, c = -6

$x = \dfrac{-5 \pm \sqrt{5^2 - 4(2)(-6)}}{2(2)}$,

$x = \dfrac{-5 \pm \sqrt{73}}{4}$,

$x = \dfrac{-5 + \sqrt{73}}{4}$, $x \approx 0.886$

$x = \dfrac{-5 - \sqrt{73}}{4}$, $x \approx -3.386$

31. $x^2 + 3x - 18 = 0$, $(x + 6)(x - 3) = 0$,

x + 6 = 0, x = -6

x - 3 = 0, x = 3

33. $3x^2 = 4x + 7$, $3x^2 - 4x - 7 = 0$,

(x + 1)(3x - 7) = 0,

x + 1 = 0, x = -1

3x - 7 = 0, 3x = 7, $x = \frac{7}{3}$, $x = 2\frac{1}{3}$

35. a = 2, b = 3, c = 5

$x = \dfrac{-3 \pm \sqrt{3^2 - 4(2)(5)}}{2(2)}$, $x = \dfrac{-3 \pm \sqrt{-31}}{4}$

No real number solution.

37. $x^2 - 6x + 9 = 0$, $(x - 3)^2 = 0$,

x - 3 = 0, x = 3

Chapter 8 Review (con't)

39.a. $A = \pi r^2, r^2 = \dfrac{A}{\pi}, \sqrt{r^2} = \sqrt{\dfrac{A}{\pi}}, r = \pm\sqrt{\dfrac{A}{\pi}}$

b. $p = \frac{1}{2}dv^2, 2p = dv^2, v^2 = \dfrac{2p}{d}$

$\sqrt{v^2} = \sqrt{\dfrac{2p}{d}}, v = \pm\sqrt{\dfrac{2p}{d}}$

c. $S = 4\pi r^2, r^2 = \dfrac{S}{4\pi}, \sqrt{r^2} = \sqrt{\dfrac{S}{4\pi}}$

$r = \pm\frac{1}{2}\sqrt{\dfrac{S}{\pi}}$

d. $h = \dfrac{v^2}{2g}, v^2 = 2gh, \sqrt{v^2} = \sqrt{2gh},$

$v = \pm\sqrt{2gh}$

41. a. To find the median put the data into ascending order:

0, 0, 0, 0, 0, 0, 0, 1, 1, 1, 2, 2, 2, 2, 2, 2, 2, 2, 2, 3, 3, 3, 3, 3, 4, 4, 4, 4, 4, 4, 5, 5, 5, 5, 5, 6, 6, 6, 6, 8, 10, 15

The median is 3.

b.

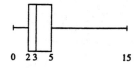

c. Mean ≈ 3.38,

standard deviation ≈ 2.92

Chapter 8 Test

1. a. $3^2 + 4^2 = 5^2$?,

$9 + 16 = 25$, right triangle

b. $1^2 + 2^2 = 3^2$?

$1 + 4 \neq 9$, not a right triangle

c. $1^2 + \left(\sqrt{3}\right)^2 = 2^2$?

$1 + 3 = 4$, right triangle

d. $\left(\sqrt{2}\right)^2 + \left(\sqrt{2}\right)^2 = 2^2$?

$2 + 2 = 4$, right triangle

e. $\left(\sqrt{3}\right)^2 + \left(\sqrt{4}\right)^2 = \left(\sqrt{5}\right)^2$?

$3 + 4 \neq 5$, not a right triangle

2. a. $\sqrt{45} = \sqrt{9 \cdot 5} = 3\sqrt{5}$

b. $\sqrt{44} = \sqrt{4 \cdot 11} = 2\sqrt{11}$

3. $\sqrt{20^2 + 5^2} = \sqrt{425} \approx 20.6$ ft.

4. a. $\sqrt{5} \cdot \sqrt{20} = \sqrt{5 \cdot 20} = \sqrt{100} = 10$

b. $\sqrt{3} \cdot \sqrt{27} = \sqrt{3 \cdot 27} = \sqrt{81} = 9$

c. $\left(3\sqrt{6}\right)^2 = 3^2\left(\sqrt{6}\right)^2 = 9 \cdot 6 = 54$

d. $\left(2\sqrt{7}\right)^2 = 2^2\left(\sqrt{7}\right)^2 = 4 \cdot 7 = 28$

e. $\sqrt{36x^2 y} = \sqrt{36}\sqrt{x^2}\sqrt{y} = 6x\sqrt{y}$

4. f. $\sqrt{0.81x^4 y^3} = \sqrt{0.81}\sqrt{(x^2)^2}\sqrt{y^2 y}$

$= 0.9x^2 y\sqrt{y}$

g. $\sqrt{\dfrac{147}{3}} = \sqrt{49} = 7$

h. $\dfrac{\sqrt{18}}{\sqrt{32}} = \sqrt{\dfrac{18}{32}} = \sqrt{\dfrac{9}{16}} = \dfrac{\sqrt{9}}{\sqrt{16}} = \dfrac{3}{4}$

i. $\sqrt{\dfrac{a^5 b^2}{a}} = \sqrt{a^4 b^2} = \sqrt{a^4}\sqrt{b^2} = a^2 b$

5. a. $a = \sqrt{\left(\sqrt{13}\right)^2 + 6^2}$, $a = \sqrt{49}$, $a = 7$

b. $\dfrac{18}{6} = \dfrac{b}{7}$, $b = 3 \cdot 7$, $b = 21$

$\dfrac{18}{6} = \dfrac{c}{\sqrt{13}}$, $c = 3\sqrt{13}$, $c \approx 10.817$

c. $\dfrac{10}{3\sqrt{13}} = \dfrac{d}{18}$, $d = \dfrac{180}{3\sqrt{13}}$, $d \approx 16.641$

$e = \sqrt{10^2 + \left(\dfrac{180}{3\sqrt{13}}\right)^2}$, $e \approx 19.415$

6. From the graph, x = -2 and x = 2.

Chapter 8 Test (con't)

7. a. $\sqrt{x^2} = \sqrt{\dfrac{36}{121}}$, $x = \pm\dfrac{6}{11}$

b. $\left(\sqrt{3-x}\right)^2 = (x+3)^2$, $3-x = x^2 + 6x + 9$

$x^2 + 7x + 6 = 0$, $(x+1)(x+6) = 0$,

$x + 1 = 0$, $x = -1$

$x + 6 = 0$, $x = -6$

$\sqrt{3-(-1)} = -1 + 3$, true

$\sqrt{3-(-6)} = -6 + 3$, false, discard

Solution is x = -1

Defined when $3 - x \geq 0$, $x \leq 3$

c. $x^2 + x - 2 = 0$, $(x + 2)(x - 1) = 0$,

$x + 2 = 0$, x = -2

$x - 1 = 0$, x = 1

d. $\left(\sqrt{5x-6}\right)^2 = 12^2$, $5x - 6 = 144$,

$5x = 150$, $x = 30$

Defined when $5x - 6 \geq 0$, $5x \geq 6$, $x \geq \frac{6}{5}$

e. $2x^2 = 8 - 15x$, $2x^2 + 15x - 8 = 0$,

$(2x - 1)(x + 8) = 0$,

$2x - 1 = 0$, $2x = 1$, $x = \frac{1}{2}$

$x + 8 = 0$, x = -8

f. $8x^2 + 5x = 4$, $8x^2 + 5x - 4 = 0$,

$a = 8$, $b = 5$, $c = -4$

$x = \dfrac{-5 \pm \sqrt{5^2 - 4(8)(-4)}}{2(8)}$,

$x = \dfrac{-5 \pm \sqrt{153}}{16}$,

$x = \dfrac{-5 + \sqrt{153}}{16}$, $x \approx 0.46$

$x = \dfrac{-5 - \sqrt{153}}{16}$, $x \approx -1.09$

g. $3x + 8 = 0$, $3x = -8$, $x = -\frac{8}{3}$

$3x - 8 = 0$, $3x = 8$, $x = \frac{8}{3}$

h. $4x^2 + 8 = 0$, $x^2 + 2 = 0$, $x^2 = -2$

no real number solution

i. $x^2 - 6x + 9 = 0$, $(x - 3)^2 = 0$,

$x - 3 = 0$, x = 3

j. $4x^2 - 25 = 0$, $(2x + 5)(2x - 5) = 0$,

$2x + 5 = 0$, $2x = -5$, $x = -\frac{5}{2}$

$2x - 5 = 0$, $2x = 5$, $x = \frac{5}{2}$

8. a. $V_e = \sqrt{\dfrac{2GM}{R}}$, $V_e^2 = \dfrac{2GM}{R}$,

$RV_e^2 = 2GM$, $R = \dfrac{2GM}{V_e^2}$

Chapter 8 Test (con't)

8. b. $E = \frac{1}{2}mv^2$, $v^2 = \dfrac{2E}{m}$, $\sqrt{v^2} = \sqrt{\dfrac{2E}{m}}$

$$v = \pm\sqrt{\dfrac{2E}{m}}$$

c. $E = \dfrac{kH^2}{8\pi}$, $H^2 = \dfrac{8\pi E}{k}$, $\sqrt{H^2} = \sqrt{\dfrac{8\pi E}{k}}$,

$$H = \pm\sqrt{\dfrac{8\pi E}{k}}, \quad H = \pm 2\sqrt{\dfrac{2\pi E}{k}}$$

9. a. $s = 8.6\sqrt{36}$, $s = 8.6 \cdot 6$, $s = 51.6\,\text{mph}$

b. $s = 8.6\sqrt{100}$, $s = 8.6 \cdot 10$, $s = 86\,\text{mph}$

c. $120 = 8.6\sqrt{t}$, $\dfrac{120}{8.6} = \sqrt{t}$, $t = \left(\dfrac{120}{8.6}\right)^2$,

$$t \approx 194.7\,\text{psi}$$

10. a. $x^2 = 50$, $\sqrt{x^2} = \sqrt{50}$, $x = 5\sqrt{2}$,

$$x \approx 7.07\,\text{ft}$$

b. $\frac{\pi}{4}x^2 = 50$, $x^2 = \dfrac{200}{\pi}$, $\sqrt{x^2} = \sqrt{\dfrac{200}{\pi}}$,

$$x = 10\sqrt{\dfrac{2}{\pi}}, \quad x \approx 7.98$$

c. $\dfrac{x^2\sqrt{3}}{4} = 50$, $x^2 = \dfrac{200}{\sqrt{3}}$, $\sqrt{x^2} = \sqrt{\dfrac{200}{\sqrt{3}}}$,

$$x = 10\sqrt{\dfrac{2}{\sqrt{3}}}, \quad x \approx 10.75$$

11. a.

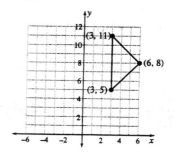

b. $\sqrt{(3-6)^2 + (5-8)^2} = \sqrt{18} \approx 4.24$

$$\sqrt{(6-3)^2 + (8-11)^2} = \sqrt{18} \approx 4.24$$
$$\sqrt{(3-3)^2 + (11-5)^2} = \sqrt{36} = 6$$

c. $\dfrac{5-8}{3-6} = 1$, $\dfrac{8-11}{6-3} = -1$, $\dfrac{11-5}{3-3} = \text{undefined}$

Two sides are perpendicular.

d. The triangle is an isosceles right triangle, it has 2 equal sides and contains a right angle.

12. a. $\dfrac{120 + 110 + 105 + 110 + 105 + 110}{6}$

$= 110$, Mean $= 110$.

$$s_x = \sqrt{\dfrac{10^2 + 0 + (-5)^2 + 0 + (-5)^2 + 0}{5}},$$

$$s_x = \sqrt{\dfrac{150}{5}}, \quad s_x = \sqrt{30}, \quad s_x \approx 5.5$$

b. Median is 145, $Q_1 = 75$, $Q_3 = 181$

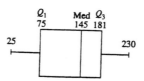

Cumulative Review Chapters 1 to 8

1. 12 hr ÷ 1.25 hr = 9

3. 0 + 3(15) - 2(20) - 10 + 25 = 20

5. $\dfrac{3x^3y^2}{27xy^3} = \dfrac{3}{27}x^{3-1}y^{2-3} = \dfrac{1}{9}x^2y^{-1} = \dfrac{x^2}{9y}$

7. 2 - 3x = 26, 2 = 26 + 3x, -24 = 3x,

x = -8

9. 3(x + 4) = 2(1 - x), 3x + 12 = 2 - 2x,

3x + 2x = 2 - 12, 5x = -10, x = -2

11. 2x - 1 ≤ 2 - x, 3x ≤ 3, x ≤ 1

13.

x	f(x) = 3x - 2x^2
-1	3(-1) - 2(-1)^2 = -5
0	3(0) - 2(0)^2 = 0
1	3(1) - 2(1)^2 = 1
2	3(2) - 2(2)^2 = -2

15. $y = \frac{4}{3}x - 2$

17.

Trips (x)	Remaining value $
0	20
2	20 - 2(2.25) = 15.50
4	20 - 4(2.25) = 11.00
6	20 - 6(2.25) = 6.50
8	20 - 8(2.25) = 2.00
10	20 - 10 (2.25) = -2.50

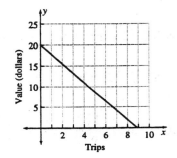

y = 20 - 2.25x

19. $\dfrac{x}{52} = \dfrac{125}{100}, \; x = \dfrac{125 \cdot 52}{100}, \; x = 65$

21. $\dfrac{33}{88} = \dfrac{x}{100}, \; x = \dfrac{3300}{88}, \; x = 37.5\%$

23. $2\sec \cdot \dfrac{100\text{ miles}}{1\text{ hour}} \cdot \dfrac{1\text{ hour}}{3600\text{ sec}} \cdot \dfrac{5280\text{ ft}}{1\text{ mile}}$

$= 293\frac{1}{3}$ ft

25. a. (x + 3)(2x - 1) = 2x^2 - x + 6x - 3

$= 2x^2 + 5x - 3$

b. x(x + 1)(3x - 1) = x(3x^2 - x + 3x - 1)

$= x(3x^2 + 2x - 1) = 3x^3 + 2x^2 - x$

Cumulative Review (con't)

27. a. $\dfrac{1.496 \times 10^8}{1.3914 \times 10^6} = \dfrac{1.496}{1.3914} \times 10^{8-6}$

$\approx 1.075 \times 10^2$

b. $\dfrac{384,000}{3480} = 110.345 \approx 1.103 \times 10^2$

c. Quotients are nearly the same.

29. $2(3x + 4y) = 2(-7),$

$6x + 8y = -14$

$\underline{+\ 7x - 8y = -25}$

$13x \qquad = -39$

$x = -3$

$3(-3) + 4y = -7, \ \ 4y = -7 + 9$

$4y = 2, \ \ y = \frac{1}{2}$

31. Let x = one leg, $x + 3$ = other leg

$x^2 + (x+3)^2 = 15^2,$
$x^2 + x^2 + 6x + 9 = 225,$
$2x^2 + 6x + 9 = 225,$
$2x^2 + 6x - 216 = 0,$
$x^2 + 3x - 108 = 0,$
$(x+12)(x-9) = 0,$
$x + 12 = 0, \ x = -12,$ discard negative
$x - 9 = 0, \ \ x = 9$
$9 + 3 = 12$

Legs are 9 and 12 units long.

33. a.

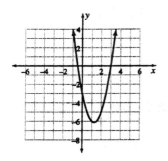

b. From the graph $x = -\frac{1}{2}$ and $x = 3$

c. $2x^2 - 5x - 3 = 0, \ \ (2x + 1)(x - 3) = 0,$

$2x + 1 = 0, \ \ 2x = -1, \ \ x = -\frac{1}{2}$

$x - 3 = 0, \ \ x = 3$

d. $a = 2, b = -5, c = -3$

$x = \dfrac{-(-5) \pm \sqrt{(-5)^2 - 4(2)(-3)}}{2(2)},$

$x = \dfrac{5 \pm \sqrt{49}}{4}, \ x = \dfrac{5 + 7}{4}, \ x = 3$

$x = \dfrac{5 - 7}{4}, \ x = -\dfrac{1}{2}$

Cumulative Review (con't)

35. a. $\dfrac{5+5+3+4+5+5+6}{7} \approx 4.7$

Mean ≈ 4.7, Median $= 5$, Mode $= 5$

b. Order data to find median, Q_1 & Q_3

0, 19, 20, 24, 26, 28, 48

Median $= 24$, $Q_1 = 19$, $Q_3 = 28$

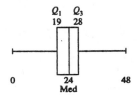

c. High data point is 348, low is 237, range is $348 - 237 = 111$

The sum of the data is 2086, there are 7 data points, mean $= \dfrac{2086}{7} = 298$

$s_x =$

$$\sqrt{\dfrac{14^2 + (-3)^2 + (-61)^2 + (-9)^2 + 4^2 + 5^2 + 50^2}{6}}$$

$$s_x = \sqrt{\dfrac{6548}{6}}, \ s_x \approx 33$$

Section 9.1

1. Expression is undefined when $x + 1 = 0$, $x = -1$, therefore it is defined for all real numbers $x \neq -1$.

3. Undefined when $2x - 1 = 0$, $x = \frac{1}{2}$, defined for all real numbers $x \neq \frac{1}{2}$

5. Undefined when $3x + 1 = 0$, $x = -\frac{1}{3}$, defined for all real numbers $x \neq -\frac{1}{3}$

7. Undefined when $x^2 + 4x + 3 = 0$, $(x + 1)(x + 3) = 0$, $x = -1$, $x = -3$, defined for all real numbers $x \neq -3$, $x \neq -1$

9. As x approaches 5 from the left, the y values of the graph approach $-\infty$. As x approaches 5 from the right, the y values approach $+\infty$.

11. As x approaches -2 from the left, the y values of the graph approach $+\infty$. As x approaches -2 from the right, the y values approach $-\infty$.

13. Solving $d = rt$ for r; $r = \dfrac{d}{t}$.

   a. $r = \dfrac{60}{2}$, $r = 30$ mph

   b. $r = \dfrac{60}{1}$, $r = 60$ mph

13. c. $r = \dfrac{60}{\frac{5}{6}}$, $r = \dfrac{60}{1} \cdot \dfrac{6}{5}$, $r = 72$ mph

   d. $r = \dfrac{60}{\frac{3}{4}}$, $r = \dfrac{60}{1} \cdot \dfrac{4}{3}$, $r = 80$ mph

   e. $r = \dfrac{60}{\frac{2}{3}}$, $r = \dfrac{60}{1} \cdot \dfrac{3}{2}$, $r = 90$ mph

   f. $r = \dfrac{60}{\frac{1}{2}}$, $r = \dfrac{60}{1} \cdot \dfrac{2}{1}$, $r = 120$ mph

15. Divide reserves by daily production for number of days, use unit analysis for number of years.

$$y = \frac{1000 \cdot 10^9}{x} \text{ days}$$

$$y = \frac{1000 \cdot 10^9}{365x} \text{ years}$$

17. Using the second equation in #15 and $x = 60,000,000$,

$$y = \frac{1000 \cdot 10^9}{365(60 \cdot 10^6)}, \quad y \approx 46 \text{ years}$$

19. a. The output is the total number of terms.

   b. If $x = 12$ the available aid will last $\dfrac{90}{12} = 7.5$ terms.

21. $r = 0$ would mean you are not moving.

Section 9.1 (con't)

23. For $t = 0$ the numerator would have to equal zero. There is no value of r that will change the numerator to zero.

25. a.

Width	Length	Area	Perimeter
1	30	30	$2(1) + 2(30)$ $= 62$
2	15	30	$2(2) + 2(15)$ $= 34$
3	10	30	$2(3) + 2(10)$ $= 26$
4	7.5	30	$2(4) + 2(7.5)$ $= 23$
5	6	30	$2(5) + 2(6)$ $= 22$
6	5	30	$2(6) + 2(5)$ $= 22$
7.5	4	30	$2(7.5) + 2(4)$ $= 23$
10	3	30	$2(10) + 2(3)$ $= 26$
15	2	30	$2(15) + 2(2)$ $= 34$
30	1	30	$2(30) + 2(1)$ $= 62$

b.

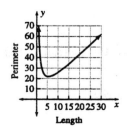

25. c. From the graph the smallest perimeter is at $\approx (5.5, 21.9)$

d. If x = length then width = $30 \div x$,

Perimeter is $y = 2x + 2(\frac{30}{x})$

27. Lines of symmetry for $y = \dfrac{1}{x}$ are $y = x$ and $y = -x$.

Line of symmetry for $y = \dfrac{1}{x^2}$ is $x = 0$.

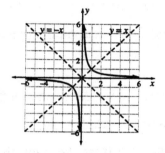

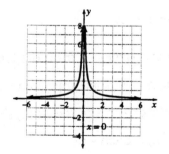

29. Yes, all points will reverse in the same manner; $y = \dfrac{1}{x}$ is equivalent to $x = \dfrac{1}{y}$.

31. The statement *varies inversely with the square of the distance* implies that d^2 will be in the denominator. The correct equation is c, $F = \dfrac{km_1m_2}{d^2}$.

Section 9.2

1. $\dfrac{2ab}{6a^2b} = \dfrac{2}{6}a^{1-2}b^{1-1} = \dfrac{1}{3}a^{-1}b^0 = \dfrac{1}{3a}$

3. $\dfrac{12cd^2}{8c^2d} = \dfrac{12}{8}c^{1-2}d^{2-1} = \dfrac{4\cdot 3}{4\cdot 2}c^{-1}d^1 = \dfrac{3d}{2c}$

5. $\dfrac{(x-2)(x+3)}{(x+3)} = \dfrac{(x-2)}{1}\cdot\dfrac{(x+3)}{(x+3)} = x-2$

7. $\dfrac{2-x}{(x+2)(x-2)} = \dfrac{1}{(x+2)}\cdot\dfrac{2-x}{(x-2)}$

   $= \dfrac{1}{x+2}\cdot\dfrac{-1(x-2)}{x-2} = \dfrac{-1}{x+2}$

9. $\dfrac{3ab+3ac}{5b^2+5bc} = \dfrac{3a(b+c)}{5b(b+c)} = \dfrac{3a}{5b}$

11. $\dfrac{4x^2+8x}{2x^2-4x} = \dfrac{4x(x+2)}{2x(x-2)} = \dfrac{2(x+2)}{x-2}$

13. $\dfrac{x^2-4}{x^2+5x+6} = \dfrac{(x+2)(x-2)}{(x+2)(x+3)} = \dfrac{x-2}{x+3}$

15. $\dfrac{x^2+x-6}{x^2-2x} = \dfrac{(x+3)(x-2)}{x(x-2)} = \dfrac{x+3}{x}$

17. $\dfrac{x-3}{6-2x} = \dfrac{(x-3)}{-2(-3+x)} = \dfrac{(x-3)}{-2(x-3)} = -\dfrac{1}{2}$

19. $\dfrac{x^2-5x+6}{x^2-9} = \dfrac{(x-2)(x-3)}{(x+3)(x-3)} = \dfrac{x-2}{x+3}$

21. $\dfrac{6}{9} = \dfrac{}{45}, \dfrac{6}{9} = \dfrac{6\cdot 5}{9\cdot 5} = \dfrac{30}{45}$

23. $\dfrac{24x}{3x^2} = \dfrac{8}{}, \dfrac{24x}{3x^2} = \dfrac{3\cdot 8\cdot x}{3\cdot x\cdot x} = \dfrac{8}{x}$

25. $\dfrac{}{10a^2} = \dfrac{b}{2a}, \dfrac{b}{2a} = \dfrac{b\cdot 5a}{2a\cdot 5a} = \dfrac{5ab}{10a^2}$

27. $\dfrac{x+2}{} = \dfrac{2x+4}{2x-6},$

   $\dfrac{2x+4}{2x-6} = \dfrac{2(x+2)}{2(x-3)} = \dfrac{x+2}{x-3}$

29. $\dfrac{a+b}{} = \dfrac{a^2+2ab+b^2}{a^2-b^2},$

   $\dfrac{a^2+2ab+b^2}{a^2-b^2} = \dfrac{(a+b)(a+b)}{(a+b)(a-b)} = \dfrac{a+b}{a-b}$

31. $\dfrac{108\,\text{m}^2}{6\,\text{m}} = \dfrac{6\cdot 18}{6}\text{m}^{2-1} = 18\,\text{m}$

33. $\dfrac{144\,\text{in}^2}{1728\,\text{in}^3} = \dfrac{144}{144\cdot 12}\text{in}^{2-3} = \dfrac{1}{12\,\text{in}}$

35. $\dfrac{2060\,\text{degree gallons}}{103\,\text{degrees}}$

   $= \dfrac{20\cdot 103\,\text{degree gallons}}{103\,\text{degrees}} = 20\,\text{gallons}$

37. $\dfrac{7-4}{4-7} = \dfrac{3}{-3} = -1$

39. $\dfrac{-3-4}{4-(-3)} = \dfrac{-7}{7} = -1$

Observe that opposite numerator and denominator simplify to -1.

41. -1(a - b) = -a + b

43. -1(-a + b) = a - b

45. -1(b + a) = -b - a

Section 9.2 (con't)

47. a. $(n - m) + (n + m) = 2n$,

   b. $-1(n - m) = -n + m$

Not opposites

49. a. $(x - 2) + (2 - x) = 0$

   b. $-1(x - 2) = -x + 2 = 2 - x$

Opposites

51. $\dfrac{x + 2}{x + 2} = 1,\ x \neq -2$

53. $-1(x - 2) = -x + 2$,

   $\dfrac{x - 2}{-x + 2} = -1,\ x \neq 2$

55. $-1(a - b) = -a + b = b - a$

   $\dfrac{b - a}{a - b} = -1,\ a \neq b$

57. $\dfrac{3 - x}{3 - x} = 1,\ x \neq 3$

59. $\frac{1}{4} = 4^{-1}$,   4 is the opposite of -4

61. $\frac{1}{a} = a^{-1}$, a is the opposite of $-a$

63. True, only factors may be canceled.

65. $\dfrac{a}{-a} = -1,\ \dfrac{-a}{a} = -1,\ -\dfrac{a}{a} = -1$

67. Fractions simplify only if the numerator and denominator contain common factors.

Section 9.3

1. a. $\dfrac{1}{3} \cdot \dfrac{1}{4} = \dfrac{1}{3 \cdot 4} = \dfrac{1}{12}$

$\dfrac{1}{3} \div \dfrac{1}{4} = \dfrac{1}{3} \cdot \dfrac{4}{1} = \dfrac{4}{3}$

b. $\dfrac{1}{2} \cdot \dfrac{1}{5} = \dfrac{1}{2 \cdot 5} = \dfrac{1}{10}$

$\dfrac{1}{2} \div \dfrac{1}{5} = \dfrac{1}{2} \cdot \dfrac{5}{1} = \dfrac{5}{2}$

3. a. $\dfrac{3}{4} \cdot \dfrac{1}{6} = \dfrac{3 \cdot 1}{4 \cdot 6} = \dfrac{3}{4 \cdot 2 \cdot 3} = \dfrac{1}{8}$

$\dfrac{3}{4} \div \dfrac{1}{6} = \dfrac{3}{4} \cdot \dfrac{6}{1} = \dfrac{3 \cdot 6}{4} = \dfrac{3 \cdot 3 \cdot 2}{2 \cdot 2} = \dfrac{9}{2}$

b. $\dfrac{2}{3} \cdot \dfrac{1}{6} = \dfrac{2 \cdot 1}{3 \cdot 6} = \dfrac{2}{3 \cdot 3 \cdot 2} = \dfrac{1}{9}$

$\dfrac{2}{3} \div \dfrac{1}{6} = \dfrac{2}{3} \cdot \dfrac{6}{1} = \dfrac{2 \cdot 6}{3 \cdot 1} = \dfrac{2 \cdot 2 \cdot 3}{3} = 4$

5. a. $100 \div \frac{4}{1} = \dfrac{100}{4} = 25$

$100 \cdot \frac{1}{4} = \dfrac{100}{1} \cdot \dfrac{1}{4} = \dfrac{100}{4} = 25$

b. $100 \div \frac{1}{5} = \dfrac{100}{1} \cdot \dfrac{5}{1} = 500$

$100 \cdot \frac{5}{1} = 500$

Observe both answers in each pair of problems are equal, division by a number is the same a multiplication by its reciprocal.

7. a. $\dfrac{1}{x} \cdot \dfrac{x^2}{1} = \dfrac{x \cdot x}{x} = x$

b. $\dfrac{1}{a} \div \dfrac{a^2 b^2}{1} = \dfrac{1}{a} \cdot \dfrac{1}{a^2 b^2} = \dfrac{1}{a^3 b^2}$

c. $\dfrac{a}{b} \cdot \dfrac{b^2}{a^2} = \dfrac{a \cdot b \cdot b}{a \cdot a \cdot b} = \dfrac{b}{a}$

d. $\dfrac{a}{b} \div \dfrac{a^2}{b^3} = \dfrac{a}{b} \cdot \dfrac{b^3}{a^2} = \dfrac{a \cdot b \cdot b \cdot b}{a \cdot a \cdot b} = \dfrac{b^2}{a}$

9. a. $\dfrac{x^2 + 2x + 1}{x + 1} \cdot \dfrac{x}{x^2 + x}$

$= \dfrac{(x+1)(x+1)x}{(x+1)x(x+1)} = 1$

b. $\dfrac{x^2 - 4}{x + 2} \cdot \dfrac{1}{x^2 - x} = \dfrac{(x+2)(x-2)}{(x+2)x(x-1)}$

$= \dfrac{x - 2}{x(x - 1)}$

c. $\dfrac{x + 2}{x^2 - 4x + 4} \div \dfrac{x^2 + 2x}{x - 2}$

$= \dfrac{x + 2}{(x-2)(x-2)} \cdot \dfrac{x - 2}{x(x + 2)} = \dfrac{1}{x(x - 2)}$

d. $\dfrac{x^2 - 5x}{x^2 + 5x} \div \dfrac{x^2 - 10x + 25}{x}$

$= \dfrac{x(x - 5)}{x(x + 5)} \cdot \dfrac{x}{(x-5)(x-5)} = \dfrac{x}{(x + 5)(x - 5)}$

Section 9.3 (con't)

9. e. $\dfrac{x^2-6x+9}{x^2+3x} \div \dfrac{x^2-9}{x}$

$= \dfrac{(x-3)(x-3)}{x(x+3)} \cdot \dfrac{x}{(x+3)(x-3)}$

$= \dfrac{x-3}{(x+3)(x+3)} = \dfrac{x-3}{(x+3)^2}$

11. a. $\dfrac{x^2+x}{x-1} \cdot \dfrac{x^2-1}{x+1} = \dfrac{x(x+1)(x+1)(x-1)}{(x-1)(x+1)}$

$= x(x+1)$

b. $\dfrac{x-3}{x^2+6x+9} \cdot \dfrac{x+3}{x^2-9}$

$= \dfrac{(x-3)(x+3)}{(x+3)(x+3)(x+3)(x-3)} = \dfrac{1}{(x+3)^2}$

c. $\dfrac{4-8x}{x+1} \div \dfrac{1-2x}{x^2-1}$

$= \dfrac{4(1-2x)}{x+1} \cdot \dfrac{(x+1)(x-1)}{1-2x} = 4(x-1)$

d. $\dfrac{x-x^2}{x+1} \cdot \dfrac{x-1}{1-x} = \dfrac{x(1-x)}{x+1} \cdot \dfrac{x-1}{1-x}$

$= \dfrac{x(x-1)}{x+1}$

e. $\dfrac{x^2-x}{x^2-3x+2} \div \dfrac{1-x^2}{x^2-2x+1}$

$= \dfrac{x(x-1)}{(x-1)(x-2)} \cdot \dfrac{(x-1)(x-1)}{(1-x)(1+x)}$

$= \dfrac{x}{x-2} \cdot (-1)\dfrac{x-1}{x+1} = -\dfrac{x(x-1)}{(x-2)(x+1)}$

13. a. $\dfrac{1}{a} \cdot \dfrac{1}{b} = \dfrac{1}{ab}$

b. $a \div \dfrac{1}{a} = a \cdot \dfrac{a}{1} = a^2$

c. $\dfrac{1}{b} \cdot a = \dfrac{a}{b}$

d. $\dfrac{1}{b} \div \dfrac{1}{a} = \dfrac{1}{b} \cdot \dfrac{a}{1} = \dfrac{a}{b}$

e. $\dfrac{a}{b} \cdot \dfrac{1}{b} = \dfrac{a}{b^2}$

f. $\dfrac{a}{b} \div \dfrac{1}{a} = \dfrac{a}{b} \cdot \dfrac{a}{1} = \dfrac{a^2}{b}$

15. $\dfrac{\dfrac{\text{miles}}{\text{miles}}}{\text{hours}} = \dfrac{\text{miles}}{1} \cdot \dfrac{\text{hours}}{\text{miles}} = \text{hours}$

17. $\dfrac{93{,}000{,}000 \text{ miles}}{186{,}000 \text{ miles per second}}$

$= \dfrac{93{,}000{,}000}{186{,}000} \cdot \dfrac{\text{miles}}{\dfrac{\text{miles}}{\text{second}}}$

$= 500\dfrac{\text{miles}}{1} \dfrac{\text{second}}{\text{miles}} = 500 \text{ seconds}$

19. $\dfrac{300 \text{ miles per hour}}{100 \text{ gallons per hour}} = \dfrac{\dfrac{300 \text{ miles}}{\text{hour}}}{\dfrac{100 \text{ gallons}}{\text{hour}}}$

$= \dfrac{300 \text{ miles}}{\text{hour}} \cdot \dfrac{\text{hour}}{100 \text{ gallons}} = \dfrac{3 \text{ miles}}{\text{gallon}}$

$= 3 \text{ mpg}$

Section 9.3 (con't)

21. $\dfrac{\dfrac{12 \text{ cookies}}{\text{dozen}}}{\dfrac{\$2.98}{\text{dozen}}} = \dfrac{12 \text{ cookies}}{\text{dozen}} \cdot \dfrac{\text{dozen}}{\$2.98}$

$= \dfrac{12 \text{ cookies}}{\$2.98} \approx 4.03 \text{ cookies per } \$$

23. $\dfrac{\dfrac{85 \text{ words}}{\text{minute}}}{\dfrac{300 \text{ words}}{\text{page}}} = \dfrac{85 \text{ words}}{\text{minute}} \cdot \dfrac{\text{page}}{300 \text{ words}}$

$\dfrac{85 \text{ page}}{300 \text{ minute}} \approx 0.28 \text{ page per minute}$

25. $\dfrac{\dfrac{40 \text{ moles}}{12 \text{ moles}}}{\text{liter}} = \dfrac{40 \text{ moles}}{1} \cdot \dfrac{\text{liter}}{12 \text{ moles}}$

$= \dfrac{40}{12} \text{ liter} = 3\tfrac{1}{3} \text{ liter}$

27. Answers will vary

29. $I = \dfrac{V}{R}, \ I = \dfrac{\dfrac{t^2 - 4}{2t^2 - 3t - 2}}{\dfrac{t+2}{t^2}}$

$I = \dfrac{(t+2)(t-2)}{(2t+1)(t-2)} \cdot \dfrac{t^2}{(t+2)}, \ I = \dfrac{t^2}{2t+1}$

31. $\dfrac{3}{5} \cdot \dfrac{4}{3} = \dfrac{3 \cdot 4}{5 \cdot 3} = \dfrac{4 \cdot 3}{5 \cdot 3} = \dfrac{4}{5}$

Mid-Chapter 9 Test

1. Undefined when $(x + 2)(x - 1) = 0$,

$x + 2 = 0$, $x = -2$

$x - 1 = 0$, $x = 1$

2.

Members	Spending	Budget
10	$\dfrac{4800}{10} = 480$	4800
20	$\dfrac{4800}{20} = 240$	4800
100	$\dfrac{4800}{100} = 48$	4800
1000	$\dfrac{4800}{1000} = 4.80$	4800
1200	$\dfrac{4800}{1200} = 4.00$	4800

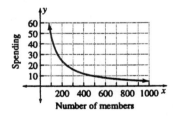

3. $y = \dfrac{4800}{x}$

4. As x approaches -2 from the left the y values approach $+\infty$.

5. As x approaches 1 from the right the y values approach $+\infty$.

6. As x approaches 1 from the left the y values approach $-\infty$.

7. a. Undefined when $28a^2 = 0$, $a = 0$,

$$\frac{24ac}{28a^2} = \frac{4 \cdot 6 \cdot a \cdot c}{4 \cdot 7 \cdot a \cdot a} = \frac{6c}{7a}$$

b. Undefined when $a - 2 = 0$, $a = 2$,

There are no common factors, expression is already simplified.

8. a. Undefined when $x^2 - 4 = 0$,

$(x - 2)(x + 2) = 0$, $x - 2 = 0$, $x = 2$

$x + 2 = 0$, $x = -2$

$$\frac{x+2}{x^2-4} = \frac{(x+2)}{(x+2)(x-2)} = \frac{1}{x-2}$$

b. Undefined when $x^2 - 3x + 2 = 0$,

$(x - 2)(x - 1) = 0$, $x - 2 = 0$, $x = 2$

$x - 1 = 0$ $x = 1$

$$\frac{x^2-2x+1}{x^2-3x+2} = \frac{(x-1)(x-1)}{(x-1)(x-2)} = \frac{x-1}{x-2}$$

9. a. $\dfrac{3x}{5y} = \dfrac{}{10xy}$, $\dfrac{3x}{5y} = \dfrac{3x \cdot 2x}{5y \cdot 2x} = \dfrac{6x^2}{10xy}$

b. $\dfrac{2a}{3b} = \dfrac{8a^2b}{}$, $\dfrac{2a}{3b} = \dfrac{2a \cdot 4ab}{3b \cdot 4ab} = \dfrac{8a^2b}{12ab^2}$

Mic-Chapter 9 Test (con't)

10. a. $\dfrac{3}{x+5} = \dfrac{}{5(x+5)}$,

$\dfrac{3}{x+5} = \dfrac{3 \cdot 5}{(x+5) \cdot 5} = \dfrac{15}{5(x+5)}$

b. $\dfrac{2}{x-3} = \dfrac{}{(x+2)(x-3)}$,

$\dfrac{2}{x-3} = \dfrac{2 \cdot (x+2)}{(x-3) \cdot (x+2)} = \dfrac{2(x+2)}{(x+2)(x-3)}$

11. $\dfrac{3x}{y^2} \div \dfrac{x^2}{y} = \dfrac{3x}{y^2} \cdot \dfrac{y}{x^2} = \dfrac{3}{xy}$

12. $\dfrac{2x}{y} \div \dfrac{x^2}{y} = \dfrac{2x}{y} \cdot \dfrac{y}{x^2} = \dfrac{2}{x}$

13. $\dfrac{x^2-5x-6}{x+2} \cdot \dfrac{x^2-4}{x-1}$

$= \dfrac{(x-6)(x+1)}{x+2} \cdot \dfrac{(x+2)(x-2)}{(x-1)}$

$= \dfrac{(x-6)(x+1)(x-2)}{x-1}$

14. $\dfrac{x^2-3x}{x^2-16} \div \dfrac{x-3}{x+4} = \dfrac{x(x-3)}{(x+4)(x-4)} \cdot \dfrac{x+4}{x-3}$

$= \dfrac{x}{x-4}$

15. $\dfrac{\dfrac{2}{3x}}{\dfrac{x^2}{6}} = \dfrac{2}{3x} \cdot \dfrac{6}{x^2} = \dfrac{4}{x^3}$

16. $\dfrac{\dfrac{63 \text{ days}}{7 \text{ days}}}{\text{week}} = \dfrac{63 \text{ days}}{1} \cdot \dfrac{\text{week}}{7 \text{ days}} = 9 \text{ weeks}$

17. $\dfrac{\dfrac{16 \text{ stitches}}{\text{second}}}{\dfrac{8 \text{ stitches}}{\text{inch}}} = \dfrac{16 \text{ stitches}}{\text{second}} \cdot \dfrac{\text{inch}}{8 \text{ stitches}}$

$= 2 \text{ inches per second}$

18. $\dfrac{\dfrac{10 \text{ mL}}{100 \text{ mL}} \cdot 400 \text{ mL}}{\dfrac{50 \text{ mL}}{100 \text{ mL}}} = \dfrac{40 \text{ mL}}{\dfrac{1}{2}}$

$= 40 \text{ mL} \cdot 2 = 80 \text{ mL}$

Section 9.4

1. **a.** $\dfrac{11}{7} - \dfrac{4}{7} = \dfrac{11-4}{7} = \dfrac{7}{7} = 1$

   b. $\dfrac{2}{3} + \dfrac{x}{3} = \dfrac{2+x}{3}$

   c. $\dfrac{3}{2z} - \dfrac{5}{2x} = \dfrac{3-5}{2x} = \dfrac{-2}{2x} = -\dfrac{1}{x}$

   d. $\dfrac{4}{x^2+1} - \dfrac{x^2}{x^2+1} = \dfrac{4-x^2}{x^2+1}$

   e. $\dfrac{2}{x-1} - \dfrac{x+1}{x-1} = \dfrac{2-(x+1)}{x-1} = \dfrac{2-x-1}{x-1}$

   $= \dfrac{-x+1}{x-1} = -1$

3. **a.** $12 = 3(4)$, $20 = 4(5)$, common denominator is $3(4)(5) = 60$

   b. Common denominator is $2x$

   c. Common denominator is y^2

   d. Common denominator is ab

   e. $x^2 - 9 = (x+3)(x-3)$, common denominator is $(x+3)(x-3)$

   f. $x^2 + 5x + 6 = (x+3)(x+2)$, $x^2 - 9 = (x+3)(x-3)$, common denominator is $(x+3)(x+2)(x-3)$

5. **a.** $\dfrac{5}{12} + \dfrac{7}{20} = \dfrac{5 \cdot 5}{12 \cdot 5} + \dfrac{7 \cdot 3}{20 \cdot 3} = \dfrac{25+21}{60}$

   $= \dfrac{46}{60} = \dfrac{2 \cdot 23}{2 \cdot 30} = \dfrac{23}{30}$

5. **b.** $\dfrac{2}{x} + \dfrac{5}{2x} = \dfrac{2 \cdot 2}{x \cdot 2} + \dfrac{5}{2x} = \dfrac{4+5}{2x} = \dfrac{9}{2x}$

   c. $\dfrac{8}{y} - \dfrac{1}{y^2} = \dfrac{8 \cdot y}{y \cdot y} - \dfrac{1}{y^2} = \dfrac{8y-1}{y^2}$

   d. $\dfrac{3}{b} + \dfrac{2}{a} + \dfrac{5}{b} - \dfrac{3}{a} = \dfrac{3a}{ab} + \dfrac{2b}{ab} + \dfrac{5a}{ab} - \dfrac{3b}{ab}$

   $\dfrac{3a+2b+5a-3b}{ab} = \dfrac{8a-b}{ab}$

   e. $\dfrac{4}{x-3} - \dfrac{2}{x^2-9}$

   $= \dfrac{4(x+3)}{(x-3)(x+3)} - \dfrac{2}{(x-3)(x+3)}$

   $= \dfrac{4x+12-2}{(x-3)(x+3)} = \dfrac{4x+10}{(x-3)(x+3)}$

   f. $\dfrac{4}{x^2+5x+6} - \dfrac{2}{x^2-9}$

   $= \dfrac{4(x-3)}{(x+3)(x+2)(x-3)} - \dfrac{2(x+2)}{(x+3)(x-3)}$

   $= \dfrac{(4x-12)-(2x+4)}{(x+3)(x+2)(x-3)}$

   $= \dfrac{4x-12-2x-4}{(x+3)(x+2)(x-3)}$

   $= \dfrac{2x-16}{(x+3)(x+2)(x-3)}$

7. $\dfrac{3}{2b} + \dfrac{3}{4a} = \dfrac{3 \cdot 2a}{2b \cdot 2a} + \dfrac{3 \cdot b}{4a \cdot b} = \dfrac{6a+3b}{4ab}$

Section 9.4 (con't)

9. $\dfrac{1}{x-1} + \dfrac{1}{x} = \dfrac{x}{x(x-1)} + \dfrac{x-1}{x(x-1)}$

$\quad = \dfrac{x+(x-1)}{x(x-1)} = \dfrac{2x-1}{x(x-1)}$

11. $\dfrac{8}{x+1} - \dfrac{3}{x} = \dfrac{8x}{x(x+1)} - \dfrac{3(x+1)}{x(x+1)}$

$\quad = \dfrac{8x-3x-3}{x(x+1)} = \dfrac{5x-3}{x(x+1)}$

13. $\dfrac{2}{x+3} + \dfrac{3}{(x+3)^2} = \dfrac{2(x+3)}{(x+3)^2} + \dfrac{3}{(x+3)^2}$

$\quad = \dfrac{2x+6+3}{(x+3)^2} = \dfrac{2x+9}{(x+3)^2}$

15. $\dfrac{1}{a} - \dfrac{1}{a^2} = \dfrac{a}{a^2} - \dfrac{1}{a^2} = \dfrac{a-1}{a^2}$

17. $\dfrac{2}{ab} - \dfrac{3}{2b} = \dfrac{2 \cdot 2}{2ab} - \dfrac{3a}{2ab} = \dfrac{4-3a}{2ab}$

19. $\dfrac{3}{x^2-x} + \dfrac{x}{x^2-3x+2}$

$\quad = \dfrac{3}{x(x-1)} + \dfrac{x}{(x-1)(x-2)}$

$\quad = \dfrac{3(x-2)}{x(x-1)(x-2)} + \dfrac{x \cdot x}{x(x-1)(x-2)}$

$\quad = \dfrac{3x-6+x^2}{x(x-1)(x-2)} = \dfrac{x^2+3x-6}{x(x-1)(x-2)}$

21. $\dfrac{T_0}{T} - 1 = \dfrac{T_0}{T} - \dfrac{T}{T} = \dfrac{T_0-T}{T}$

23. $\dfrac{1}{C_1} + \dfrac{1}{C_2} = \dfrac{C_2}{C_1 C_2} + \dfrac{C_1}{C_1 C_2}$

$\quad = \dfrac{C_2 + C_1}{C_1 C_2}$

25. $\dfrac{D}{r_1} + \dfrac{D}{r_2} = \dfrac{Dr_2}{r_1 r_2} + \dfrac{Dr_1}{r_1 r_2} = \dfrac{Dr_2 + Dr_1}{r_1 r_2}$

$\quad = \dfrac{D(r_2 + r_1)}{r_1 r_2}$

27. $\dfrac{L^2}{6d} + \dfrac{d}{2} = \dfrac{L^2}{6d} + \dfrac{3d \cdot d}{3d \cdot 2} = \dfrac{L^2 + 3d^2}{6d}$

29. $\dfrac{\frac{4}{3} - \left(-\frac{1}{3}\right)}{\frac{2}{3} - \frac{1}{3}} = \dfrac{\frac{5}{3}}{\frac{1}{3}} = \dfrac{5}{3} \cdot \dfrac{3}{1} = 5$

31. $\dfrac{\frac{1}{5} - \frac{2}{5}}{\frac{-4}{5} - \frac{1}{5}} = \dfrac{\frac{-1}{5}}{\frac{-5}{5}} = \dfrac{-1}{5} \cdot (-1) = \dfrac{1}{5}$

33. $\dfrac{\frac{a}{2} - a}{b - \frac{-b}{2}} = \dfrac{\frac{a}{2} - \frac{2a}{2}}{\frac{2b}{2} + \frac{b}{2}} = \dfrac{\frac{-a}{2}}{\frac{3b}{2}} = \dfrac{-a}{2} \cdot \dfrac{2}{3b} = -\dfrac{a}{3b}$

35. $\dfrac{0-b}{c-a} = \dfrac{-b}{c-a}$

$\dfrac{0-\frac{b}{2}}{\frac{c}{2} - \frac{a}{2}} = \dfrac{\frac{-b}{2}}{\frac{c-a}{2}} = \dfrac{-b}{2} \cdot \dfrac{2}{c-a} = \dfrac{-b}{c-a}$

Slopes are equal.

37. $\dfrac{5}{\frac{1}{5}+1} = \dfrac{5}{\frac{1}{5} + \frac{5}{5}} = \dfrac{5}{\frac{6}{5}} = \dfrac{5}{1} \cdot \dfrac{5}{6} = \dfrac{25}{6}$

Section 9.4 (con't)

39. $h = \dfrac{A}{\frac{1}{2}b}$, $h = \dfrac{A}{\frac{b}{2}}$, $h = \dfrac{A}{1} \cdot \dfrac{2}{b}$, $h = \dfrac{2A}{b}$

41. $\dfrac{x + \frac{x}{2}}{2 - \frac{x}{3}} = \dfrac{\frac{2x}{2} + \frac{x}{2}}{\frac{6}{3} - \frac{x}{3}} = \dfrac{\frac{3x}{2}}{\frac{6-x}{3}} = \dfrac{3x}{2} \cdot \dfrac{3}{6-x}$

$\quad = \dfrac{9x}{2(6-x)}$

43. $\dfrac{1}{\frac{Q_H}{Q_L} - 1} = \dfrac{1}{\frac{Q_H}{Q_L} - \frac{Q_L}{Q_L}} = \dfrac{1}{\frac{Q_H - Q_L}{Q_L}} = \dfrac{Q_L}{Q_H - Q_L}$

45. a.

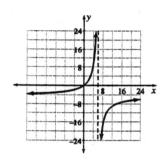

b. Graphs are the same.

c. Graph is nearly vertical near $x = 6$.

The equation is undefined at $x = 6$.

47. a. $\dfrac{3}{4} \cdot \dfrac{2}{5} = \dfrac{6}{20} = \dfrac{3}{10}$

b. $\dfrac{3}{4} \div \dfrac{2}{5} = \dfrac{3}{4} \cdot \dfrac{5}{2} = \dfrac{15}{8}$

c. $\dfrac{3}{4} + \dfrac{2}{5} = \dfrac{3 \cdot 5}{4 \cdot 5} + \dfrac{2 \cdot 4}{5 \cdot 4} = \dfrac{15 + 8}{20} = \dfrac{23}{20}$

49. a. $\dfrac{1}{a} + \dfrac{1}{b} = \dfrac{b}{ab} + \dfrac{a}{ab} = \dfrac{b + a}{ab} = \dfrac{a + b}{ab}$

b. $\dfrac{1}{a} \cdot \dfrac{1}{b} = \dfrac{1}{ab}$

Section 9.5

1. $12x\left(\dfrac{1}{12}+\dfrac{2}{3x}\right)=\dfrac{12x}{12}+\dfrac{24x}{3x}=x+8$

3. $4x^2\left(\dfrac{1}{2x}+\dfrac{3}{x^2}\right)=\dfrac{4x^2}{2x}+\dfrac{12x^2}{x^2}=2x+12$

5. $x(x-1)\left(\dfrac{1}{x-1}+\dfrac{1}{x}\right)=\dfrac{x(x-1)}{x-1}+\dfrac{x(x-1)}{x}$

      $=x+x-1=2x-1$

7. $4=2(2),\ 6=2(3),\ \text{LCD}=2(2)(3)=12$

9. LCD $= 4(5)(x) = 20x$

11. LCD $= 8x$

13. LCD $= 3x$

15. LCD $= x$

17. LCD $= (x - 1)(x + 3)$

19. LCD $= x^2$

21. LCD $= (x - 3)$

23. $\dfrac{x}{4}+\dfrac{x}{6}=28,\ 12\left(\dfrac{x}{4}+\dfrac{x}{6}\right)=12(28),$

    $3x+2x=336,\ 5x=336,\ x=67.2$

25. $20x\left(\dfrac{3}{4}+\dfrac{1}{5}\right)=20x\left(\dfrac{1}{x}\right),\ 15x+4x=20,$

    $19x=20,\ x=\dfrac{20}{19};\ x\neq 0$

27. $8x\left(\dfrac{1}{8}+\dfrac{1}{x}\right)=8x\left(\dfrac{1}{2}\right),\ x+8=4x,$

    $8=3x,\ x=\dfrac{8}{3};\ x\neq 0$

29. $3x\left(\dfrac{1}{x}\right)=3x\left(\dfrac{1}{3x}+\dfrac{1}{3}\right),\ 3=1+x,\ x=2;\ x\neq 0$

31. $\dfrac{3}{x}-\dfrac{2}{x}=\dfrac{4}{x},\ \dfrac{1}{x}=\dfrac{4}{x},\ 4x=1x,\text{no solution}$

33. $\dfrac{1}{x-1}=\dfrac{2}{x+3},\ x+3=2(x-1),$

    $x+3=2x-2,\ 5=x;\ x\neq -3,\ x\neq 1$

35. $x^2\left(\dfrac{2}{x^2}-\dfrac{3}{x}+1\right)=x^2(0),$

    $2-3x+x^2=0,\ (x-1)(x-2)=0,$
    $x-1=0,\ x=1;\ x-2=0,\ x=2;\ x\neq 0$

37. $(x-3)\left(\dfrac{1}{x-3}-3\right)=(x-3)\left(\dfrac{4-x}{x-3}\right),$

    $1-3(x-3)=4-x,\ 1-3x+9=4-x,$
    $10-3x=4-x,\ 6=2x,\ 3=x;\ x\text{ cannot}$
    equal 3, therefore no solution.

39. $\dfrac{1}{12}+\dfrac{1}{18}=\dfrac{1}{d},\ \dfrac{3}{36}+\dfrac{2}{36}=\dfrac{1}{d},\ \dfrac{5}{36}=\dfrac{1}{d},$

    $5d=36,\ d=\dfrac{36}{5}$

Section 9.5 (con't)

41. $\frac{1}{8} + \frac{1}{5} = \frac{1}{x}$, $\frac{5}{40} + \frac{8}{40} = \frac{1}{x}$, $\frac{13}{40} = \frac{1}{x}$,

$$13x = 40,\ x = \frac{40}{13}$$

43. Keystrokes will vary, $d = 7.2$

45. Keystrokes will vary, $x \approx 3.077$

47. $\frac{2}{2} + \frac{3(2)}{2} = 4?,\ 1 + 3 = 4,$

$x = 2$ is a solution

49. $\frac{2}{-\frac{1}{2}+1} + \frac{3}{-\frac{1}{2}} = -2?,$

$\frac{2}{\frac{1}{2}} - \frac{3}{\frac{1}{2}} = \frac{-1}{\frac{1}{2}} = -1 \cdot 2 = -2,$

$x = -\frac{1}{2}$ is a solution

51. $\frac{1}{14} + \frac{1}{12} = \frac{1}{d}$, $84d\left(\frac{1}{14} + \frac{1}{12}\right) = 84d\left(\frac{1}{d}\right),$

$6d + 7d = 84,\ 13d = 84,\ d = \frac{84}{13},\ d \approx 6.5\ \text{days}$

53. $\frac{1}{4} + \frac{1}{5} = \frac{1}{x}$, $20x\left(\frac{1}{4} + \frac{1}{5}\right) = 20x\left(\frac{1}{x}\right),$

$5x + 4x = 20,\ 9x = 20,\ x = \frac{20}{9},\ x \approx 2.2\ \text{hr}$

2 fans will meet the code requirement

55. $\frac{1}{5} + \frac{1}{x} = \frac{1}{3}$, $15x\left(\frac{1}{5} + \frac{1}{x}\right) = 15x\left(\frac{1}{3}\right)$

$3x + 15 = 5x,\ 15 = 2x,\ x = 7.5$

Second fan would need to vent the house in 7.5 hours.

57. $\frac{1}{9} + \frac{1}{6} = \frac{1}{x}$, $18x\left(\frac{1}{9} + \frac{1}{6}\right) = 18x\left(\frac{1}{x}\right),$

$2x + 3x = 18,\ 5x = 18,\ x = \frac{18}{5},\ x = 3.6\ \text{min}$

59. $\frac{1}{a} + \frac{1}{b} = \frac{1}{c}$, $abc\left(\frac{1}{a} + \frac{1}{b}\right) = abc\left(\frac{1}{c}\right),$

$bc + ac = ab,\ ac = ab - bc,$

$ac = b(a - c),\ \frac{ac}{a - c} = b$

61. $\frac{1}{a} + \frac{1}{b} = \frac{1}{c}$, $abc\left(\frac{1}{a} + \frac{1}{b}\right) = abc\left(\frac{1}{c}\right),$

$bc + ac = ab,\ c(b + a) = ab,$

$c = \frac{ab}{b + a}$

63. Answers will vary.

65. $x \neq 0;$

$5x + \frac{13}{2} = \frac{3}{2x}$, $2x\left(5x + \frac{13}{2}\right) = 2x\left(\frac{3}{2x}\right),$

$10x^2 + 13x = 3,\ 10x^2 + 13x - 3 = 0,$

$(5x - 1)(2x + 3) = 0;\ 5x - 1 = 0,\ 5x = 1,\ x = \frac{1}{5};$

$2x + 3 = 0,\ 2x = -3,\ x = -\frac{3}{2}$

Section 9.5 (con't)

67. x ≠ 0, x ≠ -1;

$$\frac{1}{x} + 3 = \frac{5}{x(x+1)},$$

$$x(x+1)\left(\frac{1}{x} + 3\right) = x(x+1)\left(\frac{5}{x(x+1)}\right),$$

$$x + 1 + 3x(x+1) = 5, \; x + 1 + 3x^2 + 3x = 5,$$

$$3x^2 + 4x + 1 = 5, \; 3x^2 + 4x - 4 = 0,$$

$$(3x-2)(x+2) = 0; \; x + 2 = 0, \; x = -2;$$

$$3x - 2 = 0, \; 3x = 2, \; x = \tfrac{2}{3};$$

69. x ≠ 1;

$$x = 6 - \frac{6}{x-1},$$

$$(x-1)(x) = (x-1)\left(6 - \frac{6}{x-1}\right),$$

$$x^2 - x = 6(x-1) - 6, \; x^2 - x = 6x - 6 - 6,$$

$$x^2 - x = 6x - 12, \; x^2 - 7x + 12 = 0,$$

$$(x-3)(x-4) = 0; \; x - 3 = 0, \; x = 3;$$

$$x - 4 = 0, \; x = 4$$

71. x ≠ 2;

$$\frac{x+3}{x-2} = \frac{1}{x-2},$$

$$(x+3)(x-2) = x - 2,$$

$$x^2 - 2x + 3x - 6 = x - 2,$$

$$x^2 + x - 6 = x - 2, \; x^2 - 4 = 0,$$

$$(x+2)(x-2) = 0; \; x + 2 = 0, \; x = -2$$

$$x - 2 = 0, \; x = 2 \text{ - discard this answer}$$

Chapter 9 Review

1. Undefined when x + 3 = 0, x = -3

3. $y = \dfrac{4000 \text{ trillion ft}^3}{\dfrac{\text{x ft}^3}{\text{day}} \cdot \dfrac{365 \text{ day}}{\text{year}}} = \dfrac{4000 \text{ trillion}}{365x}$ years

5. a. As x approaches -1 from the right the y values approach $-\infty$.

 b. As x approaches -1 from the left the y values approach $-\infty$.

7. $\dfrac{2xy}{x^2} = \dfrac{2y}{x}$

9. $\dfrac{a^2 - b^2}{a + b} = \dfrac{(a+b)(a-b)}{a+b} = a - b$

11. No common factors, expression is already simplified.

13. $\dfrac{1-a}{a-1} = -1; a \neq 1$

15. $-1(x - 3) = -x + 3 = 3 - x$

 $\dfrac{x-3}{3-x} = -1$

17. $\dfrac{a-b}{a-b} = 1$

19. $-1(4 - x) = -4 + x = x - 4$

 $\dfrac{x-4}{4-x} = -1$

21. $\dfrac{16}{9} = \dfrac{\sqrt{16}}{\sqrt{9}}?, \dfrac{4 \cdot 4}{3 \cdot 3} \neq \dfrac{4}{3}$

 Fractions are not equal.

23. a. $300 \div \frac{3}{1} = \dfrac{300}{3} = 100$

 $300 \cdot \frac{1}{3} = \dfrac{300}{1} \cdot \dfrac{1}{3} = \dfrac{300}{3} = 100$

 b. Observe answers are the same.

 c. Division by a number is the same as multiplication by its reciprocal.

25. Phrase *each person eats one-third* indicates division.

 $\dfrac{4 \text{ pizzas}}{\frac{1}{3} \text{ pizza per serving}} = 4 \cdot \frac{3}{1} \text{ servings}$

 $= 12 \text{ servings}$

27. The phrase *half as far* indicates either division by 2 or multiplication by one-half.

 $\dfrac{3}{4} \cdot \dfrac{1}{2} = \dfrac{3}{8} \text{ mile}$

29. The phrase *what fraction remains* indicates subtraction.

 $1 - \frac{1}{2} - \frac{1}{3} = \dfrac{6}{6} - \dfrac{3}{6} - \dfrac{2}{6} = \dfrac{6-3-2}{6} = \dfrac{1}{6}$

31. $\dfrac{1-x}{x+1} \div \dfrac{x^2 - 1}{x^2} = \dfrac{1-x}{x+1} \cdot \dfrac{x^2}{(x-1)(x+1)}$

 $= \dfrac{-1(x-1)}{x+1} \cdot \dfrac{x^2}{(x-1)(x+1)} = \dfrac{-x^2}{(x+1)^2}$

Chapter 9 Review (con't)

33. $\dfrac{n-2}{n(n-1)} \cdot \dfrac{(n+1)n(n-1)}{n-2} = n+1$

35. $\dfrac{3x-9}{x+3} \cdot \dfrac{x}{x^2-6x+9}$

$= \dfrac{3(x-3)}{x+3} \cdot \dfrac{x}{(x-3)(x-3)} = \dfrac{3x}{(x+3)(x-3)}$

37. $\dfrac{x}{x+2} - \dfrac{2}{x^2-4} = \dfrac{x}{x+2} - \dfrac{2}{(x+2)(x-2)}$

$= \dfrac{x(x-2)-2}{(x+2)(x-2)} = \dfrac{x^2-2x-2}{(x+2)(x-2)}$

39. $\dfrac{a}{a+b} - \dfrac{b}{a-b} = \dfrac{a(a-b)-b(a+b)}{(a+b)(a-b)}$

$= \dfrac{a^2-ab-ab-b^2}{(a+b)(a-b)} = \dfrac{a^2-2ab-b^2}{(a+b)(a-b)}$

41. $1 - \dfrac{x^2}{2} + \dfrac{x^4}{24} - \dfrac{x^6}{720}$

$= \dfrac{720 - 360x^2 + 30x^4 - x^6}{720}$

43. $\dfrac{1}{a} + \dfrac{2a}{3} \cdot \dfrac{6}{a} \div \dfrac{1}{3} - \dfrac{a}{3} = \dfrac{1}{a} + \dfrac{12a}{3a} \cdot \dfrac{3}{1} - \dfrac{a}{3}$

$= \dfrac{1}{a} + \dfrac{36a}{3a} - \dfrac{a}{3} = \dfrac{3}{3a} + \dfrac{36a}{3a} - \dfrac{a^2}{3a}$

$= \dfrac{3 + 36a - a^2}{3a}$

45. $\dfrac{4 \text{ buttons}}{\text{card}} \div \dfrac{12 \text{ buttons}}{\text{shirt}}$

$= \dfrac{4 \text{ buttons}}{\text{card}} \cdot \dfrac{\text{shirt}}{12 \text{ buttons}} = \dfrac{1 \text{ shirt}}{3 \text{ card}}$

$= \dfrac{1}{3} \text{ shirt per card}$

47. $\text{gr } \dfrac{1}{2} \cdot \dfrac{1 \text{ tab}}{\text{gr } \frac{1}{6}} = \dfrac{1}{2} \cdot 1 \text{ tab} \cdot \dfrac{6}{1} = 3 \text{ tab}$

49. $\text{gr } \dfrac{1}{150} \cdot \dfrac{1 \text{ mL}}{\text{gr } \frac{1}{750}} = \dfrac{1}{150} \cdot 1 \text{ mL} \cdot \dfrac{750}{1} = 5 \text{ mL}$

51. $t^2 = \dfrac{d}{\frac{1}{2}g},\; t^2 = \dfrac{d}{g} \cdot \dfrac{2}{1},\; t^2 = \dfrac{2d}{g}$

53. $(x+2)(x-2)\left(\dfrac{2}{x-2} + \dfrac{1}{x+2}\right)$

$= 2(x+2) + 1(x-2) = 2x+4+x-2$

$= 3x+2$

55. $\dfrac{1}{5} + \dfrac{1}{x} = \dfrac{8}{5x},\; 5x\left(\dfrac{1}{5} + \dfrac{1}{x}\right) = 5x\left(\dfrac{8}{5x}\right),$

$x+5 = 8,\; x = 3;\; x \neq 0$

57. $\dfrac{1}{2x} = \dfrac{2}{x} - \dfrac{x}{24},\; 24x\left(\dfrac{1}{2x}\right) = 24x\left(\dfrac{2}{x} - \dfrac{x}{24}\right)$

$12 = 48 - x^2,\; x^2 = 36,\; \sqrt{x^2} = \sqrt{36},$
$x = \pm 6;\; x \neq 0$

59. $\dfrac{1}{F} = \dfrac{1}{f_1} + \dfrac{1}{f_2},\; \dfrac{1}{F} = \dfrac{f_2}{f_1 f_2} + \dfrac{f_1}{f_1 f_2}$

$\dfrac{1}{F} = \dfrac{f_2 + f_1}{f_1 f_2}$

Chapter 9 Review (con't)

61. $\dfrac{1}{3}+\dfrac{1}{2}=\dfrac{1}{x}, \ 6x\left(\dfrac{1}{3}+\dfrac{1}{2}\right)=6x\left(\dfrac{1}{x}\right)$

$2x+3x=6, \ 5x=6, \ x=\dfrac{6}{5}, \ x=1.2 \text{ min}$

Chapter 9 Test

1. Undefined for x - 4 = 0, x = 4

2. Expression simplifies to -1 because 4 - x is the opposite of x - 4.

3.

x	$y = \dfrac{4}{x-2}$
-5	$\dfrac{4}{-5-2} = \dfrac{4}{-7}$
-4	$\dfrac{4}{-4-2} = \dfrac{4}{-6} = -\dfrac{2}{3}$
-3	$\dfrac{4}{-3-2} = \dfrac{4}{-5}$
-2	$\dfrac{4}{-2-2} = \dfrac{4}{-4} = -1$
-1	$\dfrac{4}{-1-2} = \dfrac{4}{-3} = -1\dfrac{1}{3}$
0	$\dfrac{4}{0-2} = \dfrac{4}{-2} = -2$
1	$\dfrac{4}{1-2} = \dfrac{4}{-1} = -4$
2	$\dfrac{4}{2-2}$, undefined
3	$\dfrac{4}{3-2} = \dfrac{4}{1} = 4$
4	$\dfrac{4}{4-2} = \dfrac{4}{2} = 2$
5	$\dfrac{4}{5-2} = \dfrac{4}{3} = 1\dfrac{1}{3}$

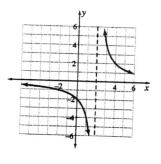

4. $\dfrac{6}{9} = \dfrac{2 \cdot 3}{3 \cdot 3} = \dfrac{2}{3}$

   a. $\dfrac{6 \div 3}{9 \div 3} = \dfrac{2}{3}$, equivalent

   b. $\dfrac{6 - 3}{9 - 3} = \dfrac{3}{6} = \dfrac{1}{2}$

   c. $\dfrac{6 + 3}{9 + 3} = \dfrac{9}{12} = \dfrac{3 \cdot 3}{4 \cdot 3} = \dfrac{3}{4}$

   d. $\dfrac{6 \cdot 3}{9 \cdot 3} = \dfrac{6}{9} = \dfrac{2}{3}$, equivalent

   Only a and d satisfy the simplification properties of fractions

5. $\dfrac{4}{25} = \dfrac{2 \cdot 2}{5 \cdot 5}, \dfrac{\sqrt{4}}{\sqrt{25}} = \dfrac{2}{5}, \dfrac{2 \cdot 2}{5 \cdot 5} \neq \dfrac{2}{5}$

6. $\dfrac{b^2}{3ab} = \dfrac{b}{3a}$

7. The expression has no common factors, it is already simplified.

8. $\dfrac{a^2 - b^2}{a - b} = \dfrac{(a-b)(a+b)}{a-b} = a + b$

9. $\dfrac{xy}{xy + y} = \dfrac{xy}{y(x+1)} = \dfrac{x}{x+1}$

10. $\dfrac{1-x}{x+1} \cdot \dfrac{(x+1)(x-1)}{x^2} = \dfrac{(1-x)(x-1)}{x^2}$

   or $\dfrac{-1(x-1)(x-1)}{x^2} = -\dfrac{(x-1)^2}{x^2}$

Chapter 9 Test (con't)

11. $\dfrac{x-1}{(1-x)(1+x)} \div \dfrac{(x-1)(x-1)}{1+x}$

$\quad = \dfrac{x-1}{(1-x)(1+x)} \cdot \dfrac{1+x}{(x-1)(x-1)}$

$\quad = \dfrac{1}{(1-x)(x-1)},$

$\quad or\ \dfrac{1}{-1(x-1)(x-1)} = -\dfrac{1}{(x-1)^2}$

12. $\dfrac{n+1}{n} \div \dfrac{n(n-1)}{n^2} = \dfrac{n+1}{n} \cdot \dfrac{n^2}{n(n-1)}$

$\quad = \dfrac{n+1}{n-1}$

13. $\dfrac{x(x-2)}{(x-2)(x+2)} \cdot \dfrac{x-2}{x} = \dfrac{x-2}{x+2}$

14. $(x+1)(x-1)\left(\dfrac{1}{x+1} + \dfrac{2}{x-1}\right)$

$\quad = 1(x-1) + 2(x+1) = x-1+2x+2 = 3x+1$

15. $\dfrac{1}{16} + \dfrac{1}{x} = \dfrac{1}{5},\ 80x\left(\dfrac{1}{16} + \dfrac{1}{x}\right) = 80x\left(\dfrac{1}{5}\right)$

$\quad 5x + 80 = 16x,\ 80 = 11x,\ \dfrac{80}{11} = x$

$\quad x \approx 7.3\ \text{min}$

16. $\dfrac{4\ \text{yd}}{\text{shirt}} \cdot \dfrac{\$8.98}{\text{yd}} = \dfrac{\$35.92}{\text{shirt}}$

17. $\dfrac{\dfrac{\$2.50}{1\ \text{gal}}}{\dfrac{25\ \text{mi}}{1\ \text{gal}}} = \dfrac{\$2.50}{1\ \text{gal}} \cdot \dfrac{1\ \text{gal}}{25\ \text{mi}} = \$0.10\ \text{per mi}$

18. $3\ \text{mg} \cdot \dfrac{\text{gr}\,1}{60\ \text{mg}} \cdot \dfrac{1\ \text{tab}}{\text{gr}\,\frac{1}{120}} = \dfrac{3\ \text{tab}}{60} \cdot \dfrac{120}{1} = 6\ \text{tab}$

19. $\dfrac{2\frac{1}{7}}{10} = \dfrac{3}{14},\ 14(2\frac{1}{7}) = 3(10),\ 14(\frac{15}{7}) = 30,$

$\quad 2(15) = 30,\ 30 = 30$

20. $\dfrac{1}{p} + \dfrac{1}{q} = \dfrac{q}{pq} + \dfrac{p}{pq} = \dfrac{q+p}{pq}$

21. $\dfrac{A}{\frac{1}{2}b} = \dfrac{A}{b} \cdot \dfrac{2}{1} = \dfrac{2A}{b}$

22. $\dfrac{2}{9} + \dfrac{7}{15} = \dfrac{2 \cdot 5}{9 \cdot 5} + \dfrac{7 \cdot 3}{15 \cdot 3} = \dfrac{10+21}{45} = \dfrac{31}{45}$

23. $\dfrac{2}{ab^2} - \dfrac{a}{2b} = \dfrac{2(2)}{ab^2(2)} - \dfrac{a(ab)}{2b(ab)} = \dfrac{4-a^2b}{2ab^2}$

24. $\dfrac{x}{x+2} + \dfrac{2}{(x+2)(x-2)} = \dfrac{x(x-2)+2}{(x+2)(x-2)}$

$\quad = \dfrac{x^2-2x+2}{(x+2)(x-2)}$

25. $\dfrac{2}{3} + \dfrac{3}{x} - \dfrac{4}{x^2} = \dfrac{2(x^2)}{3(x^2)} + \dfrac{3(3x)}{x(3x)} - \dfrac{4(3)}{x^2(3)}$

$\quad = \dfrac{2x^2+9x-12}{3x^2}$

Chapter 9 Test (con't)

26. $\dfrac{a}{2} + \dfrac{2}{3} \cdot \dfrac{a}{2} - \dfrac{3a}{2} \div \dfrac{9}{2} + \dfrac{1}{4}$

$$= \dfrac{a}{2} + \dfrac{2a}{6} - \dfrac{3a}{2} \cdot \dfrac{2}{9} + \dfrac{1}{4}$$

$$= \dfrac{a}{2} + \dfrac{2a}{6} - \dfrac{6a}{18} + \dfrac{1}{4}$$

$$= \dfrac{a}{2} + \dfrac{a}{3} - \dfrac{a}{3} + \dfrac{1}{4} = \dfrac{a}{2} + \dfrac{1}{4}$$

$$= \dfrac{2a+1}{4}$$

27. $\dfrac{1}{3} + \dfrac{1}{6} = \dfrac{1}{x}, \ 6x\left(\dfrac{1}{3} + \dfrac{1}{6}\right) = 6x\left(\dfrac{1}{x}\right)$

$2x + x = 6, \ 3x = 6, \ x = 2; \ x \neq 0$

28. $\dfrac{1}{x} = \dfrac{1}{2x} + \dfrac{1}{2}, \ 2x\left(\dfrac{1}{x}\right) = 2x\left(\dfrac{1}{2x} + \dfrac{1}{2}\right)$

$2 = 1 + x, \ x = 1; \ x \neq 0$

29. $\dfrac{3}{x} = \dfrac{2}{x-3}, \ 3(x-3) = 2x,$

$3x - 9 = 2x, \ x = 9; \ x \neq 0, \ x \neq 3$

30. $\dfrac{x-2}{4} + \dfrac{1}{x} = \dfrac{19}{4x},$

$$4x\left(\dfrac{x-2}{4} + \dfrac{1}{x}\right) = 4x\left(\dfrac{19}{4x}\right),$$

$x(x-2) + 4 = 19, \ x^2 - 2x + 4 = 19,$

$x^2 - 2x - 15 = 0, \ (x-5)(x+3) = 0;$

$x - 5 = 0, \ x = 5;$

$x + 3 = 0, \ x = -3; \ x \neq 0$

31. Factoring permits us to simplify expressions before multiplying. After changing division to multiplication we simplify by factoring.

Final Exam Review

1. **a.** $-7 + (-5) = -12$

b. $-3 - (-8) = -3 + (+8) = 5$

c. $-6 + 11 = 5$

d. $(-5)(12) = -60$

e. $(-4)(-16) = 64$

f. $|4 + (-8)| = |-4| = 4$

g. $\left|\dfrac{-45}{9}\right| = |-5| = 5$

h. $\dfrac{3}{2} \cdot \dfrac{4}{15} = \dfrac{3}{2} \cdot \dfrac{2 \cdot 2}{3 \cdot 5} = \dfrac{2}{5}$

i. $\dfrac{-2}{3} \div \dfrac{5}{6} = \dfrac{-2}{3} \cdot \dfrac{6}{5} = \dfrac{-2}{3} \cdot \dfrac{2 \cdot 3}{5} = \dfrac{-4}{5}$

3. **a.** $\dfrac{bcd}{bdf} = \dfrac{c}{f}$

b. $\dfrac{-3rs}{12sx} = \dfrac{-3rs}{3 \cdot 4sx} = \dfrac{-r}{4x}$

c. $\dfrac{a^3}{a^2} = a^{3-2} = a^1 = a$

d. $\dfrac{x-2}{2-x} = -1$

e. $\dfrac{(4m^2n)^3}{6n^2} = \dfrac{4^3 m^{2\cdot3} n^3}{6n^2} = \dfrac{64}{6} m^6 n^{3-2}$

$= \dfrac{2 \cdot 32}{2 \cdot 3} m^6 n^1 = \dfrac{32 m^6 n}{3}$

f. $\dfrac{ab}{c} \div \dfrac{ac}{b} = \dfrac{ab}{c} \cdot \dfrac{b}{ac} = \dfrac{b^2}{c^2}$

3. **g.** $n^3 n^4 = n^{3+4} = n^7$

h. $(mn^2)^3 = m^3 n^{2\cdot3} = m^3 n^6$

i. $(m^{-1} m^{-2}) = m^{-1-2} = m^{-3} = \dfrac{1}{m^3}$

j. $(3x^3)^0 = 1$

k. $\dfrac{6x^{-3}}{2x^2 y^{-2}} = \dfrac{6}{2} x^{-3-2} y^{-(-2)}$

$= 3x^{-5} y^2 = \dfrac{3y^2}{x^5}$

l. $(3x^{-3} y^2)^3 = 3^3 x^{-3\cdot3} y^{2\cdot3} = \dfrac{27 y^6}{x^9}$

5. **a.** $2x + 3y = 12,\ 3y = 12 - 2x,$

$y = \dfrac{12 - 2x}{3},\ y = 4 - \tfrac{2}{3}x$

b. $2x - 3y = 7,\ 2x = 7 + 3y$

$x = \dfrac{7 + 3y}{2},\ x = \tfrac{7}{2} + \tfrac{3}{2}y$

c. $ax + by = c,\ by = c - ax,$

$y = \dfrac{c - ax}{b},\ y = \tfrac{c}{b} - \tfrac{a}{b}x$

d. $\dfrac{x}{x+1} = \dfrac{3}{8},\ 8x = 3(x+1),\ 8x = 3x + 3,$

$5x = 3,\ x = \tfrac{3}{5}$

Final Exam Review (con't)

5. e. $\dfrac{P_1 V_1}{T_1} = \dfrac{P_2 V_2}{T_2},$

$\dfrac{T_2}{V_2}\left(\dfrac{P_1 V_1}{T_1}\right) = \dfrac{T_2}{V_2}\left(\dfrac{P_2 V_2}{T_2}\right),$

$\dfrac{P_1 V_1 T_2}{T_1 V_2} = P_2$

f. $3(x+2) = 7(x-10),$

$3x + 6 = 7x - 70,\ 76 = 4x,\ x = 19$

g. $4^2 + x^2 = 8^2,\ x^2 = 64 - 16$

$\sqrt{x^2} = \sqrt{48},\ x = \pm\sqrt{48},\ x = \pm\sqrt{16 \cdot 3}$

$x = \pm 4\sqrt{3},\ x \approx \pm 6.9$

h. $d^2 = [4 - (-3)]^2 + (5 - 2)^2,$

$d^2 = (4+3)^2 + 3^2,\ d^2 = 7^2 + 3^2,$

$d^2 = 49 + 9,\ d^2 = 58,\ \sqrt{d^2} = \sqrt{58},$

$d = \pm\sqrt{58},\ d \approx \pm 7.6$

i. $6 = \sqrt{3x-3},\ 6^2 = \left(\sqrt{3x-3}\right)^2,$

$36 = 3x - 3,\ 39 = 3x,\ 13 = x$

7. a. $12xy + 3xy^2 + 6x^2 y^2$

$= 3xy(4) + 3xy(y) + 3xy(2xy)$

$= 3xy(4 + y + 2xy)$

b. $x^2 + x = x \cdot x + x \cdot 1 = x(x+1)$

7. c.

factor	x	$+5$
x	x^2	$5x$
-4	$-4x$	-20

$x^2 + x - 20 = (x+5)(x-4)$

d. $4x^2 - 9 = (2x+3)(2x-3)$

e.

factor	$2x$	-1
$3x$	$6x^2$	$-3x$
$+2$	$4x$	-2

$6x^2 + x - 2 = (2x-1)(3x+2)$

f.

factor	x	-2
$6x$	$6x^2$	$-12x$
$+1$	x	-2

$6x^2 - 11x - 2 = (x-2)(6x+1)$

g. $x^2 - 10x + 25 = (x-5)^2$

h.

factor	x	-1
$6x$	$6x^2$	$-6x$
$+9$	$9x$	-9

$6x^2 + 3x - 9 = (x-1)(6x+9)$

9. From the graph solution is (1, -1).

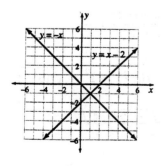

Final Exam Review (con't)

11. $3x + 4y = 10$, $x = 6 - 2y$

$3(6 - 2y) + 4y = 10$,

$18 - 6y + 4y = 10$,

$-2y = -8$, $y = 4$

$x = 6 - 2(4)$, $x = -2$

13. $2(3x + 4y) = 2(-20)$

$-3(2x - 3y) = -3(-2)$

$6x + 8y = -40$

$+ \underline{-6x + 9y = 6}$

$17y = -34$, $y = -2$

$3x + 4(-2) = -20$,

$3x - 8 = -20$, $3x = -12$,

$x = -4$

15. $x + y = 8$, $x = 8 - y$

$(8 - y) + y = 8$, $8 = 8$

always true, infinite number of solutions

17. slope: $\dfrac{5 - (-3)}{2 - 2} = \dfrac{8}{0}$, undefined

distance: $\sqrt{0^2 + 8^2} = \sqrt{8^2} = 8$

19. slope: $\dfrac{-2 - (-3)}{-5 - 2} = \dfrac{1}{-7}$

distance: $\sqrt{(-7)^2 + 1^2} = \sqrt{49 + 1}$

$= \sqrt{50} = \sqrt{25 \cdot 2} = 5\sqrt{2}$

21. $\dfrac{4}{x} = \dfrac{6}{x}$, $4x = 6x$, always false, no solution

23. $\dfrac{2x + 26}{3} = \dfrac{x - 1}{2}$, $2(2x + 26) = 3(x - 1)$

$4x + 52 = 3x - 3$, $x = -55$

25. $\dfrac{x}{1} = \dfrac{14}{x + 5}$, $x(x + 5) = 14$, $x^2 + 5x = 14$,

$x^2 + 5x - 14 = 0$, $(x + 7)(x - 2) = 0$,

$x + 7 = 0$, $x = -7$

$x - 2 = 0$, $x = 2$

27. $\dfrac{2x}{5} - \dfrac{x}{10} = \dfrac{3x}{10}$, $\dfrac{4x}{10} - \dfrac{x}{10} = \dfrac{3x}{10}$, $\dfrac{3x}{10} = \dfrac{3x}{10}$

always true, infinite number of solutions

29. $\sqrt{x + 13} = 5$, $\left(\sqrt{x + 13}\right)^2 = 5^2$,

$x + 13 = 25$, $x = 12$

31. $\sqrt{x + 13} = 0$, only if $x + 13 = 0$, $x = -13$

33. a. From the graph $(4, 0)$

b. From the graph $(0, 2)$

c. From the graph $x = 3$ when $y = 1$.

d. There is no value for y when $x = 13$, y is undefined when $x = 13$.

e. There are no negative values for y because $\sqrt{4 - x}$ is always positive.

35. The dieter went off the diet between the 10th and 15th day and gained weight.

Final Exam Review (con't)

37. a. $3^0 + 4^2 = 1 + 16 = 17$

b. $5^2 + 5^{-1} = 25 + \frac{1}{5} = 25.2$

c. $\sqrt{25} + 16^{\frac{1}{2}} = \sqrt{25} + \sqrt{16} = 5 + 4 = 9$

d. $\sqrt{225} + 25^{\frac{1}{2}} = \sqrt{225} + \sqrt{25}$

$= 15 + 5 = 20$

e. $\sqrt{64x^2} = \sqrt{64}\sqrt{x^2} = 8|x|$

f. $\sqrt{64x^2} = \sqrt{64}\sqrt{x^2} = 8x$

g. 0^0 is undefined

h. $\sqrt{x}$ is undefined if $x < 0$

39. $\dfrac{4x^2 - 1}{2x^2 + 5x + 2} = \dfrac{(2x+1)(2x-1)}{(2x+1)(x+2)}$

$= \dfrac{2x - 1}{x + 2}$

41. a. $\dfrac{\$6.00}{hour} \div \dfrac{3\ \text{rooms cleaned}}{hour}$

$= \dfrac{\$6.00}{hour} \cdot \dfrac{hour}{3\ \text{rooms cleaned}}$

$= \$2.00$ per room cleaned

41. b. $d = -\dfrac{1}{2}\left(\dfrac{32.2\,\text{ft}}{\text{sec}^2}\right)(3\,\text{sec})^2$

$+ \left(\dfrac{22\,\text{ft}}{\text{sec}}\right)(3\,\text{sec}) + 100\,\text{ft},$

$d = -\dfrac{(32.2\,\text{ft})3^2}{2} + (22\,\text{ft})3 + 100\,\text{ft}$

$d = -16.1(9)\,\text{ft} + 66\,\text{ft} + 100\,\text{ft}$

$d = (-144.9 + 66 + 100)\,\text{ft},\ d = 21.1\,\text{ft}$

43. a. $\dfrac{2 - x}{xy^2} \cdot \dfrac{x^2}{x - 2} = \dfrac{-1(x - 2)}{xy^2} \cdot \dfrac{x^2}{x - 2},$

$= -\dfrac{x}{y^2}$

b. $\dfrac{(x - 4)(x + 1)}{x(x - 2)} \cdot \dfrac{(x + 2)(x - 2)}{(-1)(x - 4)}$

$= -\dfrac{(x + 1)(x + 2)}{x}$

c. $\dfrac{x - 4}{x + 2} + \dfrac{(x - 4)(x + 1)}{(x - 2)(x + 2)}$

$= \dfrac{(x - 2)(x - 4)}{(x - 2)(x + 2)} + \dfrac{(x - 4)(x + 1)}{(x - 2)(x + 2)},$

$= \dfrac{x^2 - 6x + 8 + x^2 - 3x - 4}{(x - 2)(x + 2)},$

$= \dfrac{2x^2 - 9x + 4}{(x - 2)(x + 2)} = \dfrac{(2x - 1)(x - 4)}{(x - 2)(x + 2)}$

d. $\dfrac{(x + 2)(x + 1)}{(x + 3)(x + 3)} \div \dfrac{x + 2}{x + 3}$

$= \dfrac{(x + 2)(x + 1)}{(x + 3)(x + 3)} \cdot \dfrac{(x + 3)}{(x + 2)} = \dfrac{x + 1}{x + 3}$

Final Exam Review (con't)

43. e. $\dfrac{1}{3x} - \dfrac{2}{5x} = \dfrac{1 \cdot 5}{3x \cdot 5} - \dfrac{2 \cdot 3}{5x \cdot 3}$

$= \dfrac{5}{15x} - \dfrac{6}{15x} = -\dfrac{1}{15x}$

f. $8x\left(\dfrac{1}{2x} - \dfrac{3}{x}\right) = 4 - 24 = -20$

g. $3(x-2)\left(\dfrac{4}{x-2} + \dfrac{x}{3}\right) = 12 + x(x-2)$

$= 12 + x^2 - 2x = x^2 - 2x + 12$

45. a

Balance	Payment
$25	$25
$35	30 + 0.2(35 - 30) = $31
$95	30 + 0.20(95 - 30) = $43
$100	30 + 0.20(100 - 30) = $44
$105	50 + 0.50(105 - 100) = $52.50

b. For $(0, 30]$ $y = x$

For $(30, 100]$ $y = 30 + 0.20(x - 30)$

For $(100, +\infty)$ $y = 50 + 0.50(x - 100)$

c. [ ] includes endpoints, ( ) excludes endpoints

d. Yes, it is not possible to include ∞ as an endpoint.

e. $(0, 30]$ is $0 < x \le 30$

$(30, 100]$ is $30 < x \le 100$

$(100, +\infty)$ is $x > 100$

47. a. $2756.4 \times 10^6 \approx 2.76 \times 10^9$ miles

b. $\dfrac{4551.4 \times 10^6 \text{ miles}}{1} \cdot \dfrac{1 \sec}{186,000 \text{ mi}}$

$\approx \dfrac{24469.9 \sec}{1} \cdot \dfrac{1 \min}{60 \sec} \cdot \dfrac{1 \text{ hr}}{60 \min} \approx 6.8 \text{ hr}$

c. $\dfrac{200 \text{ m}}{125.96 \sec} \cdot \dfrac{1 \text{ mi}}{1609 \text{ m}} \cdot \dfrac{60 \sec}{1 \min} \cdot \dfrac{60 \min}{1 \text{ hr}}$

$\approx 3.55 \text{ mph}$

d. $\dfrac{28,300 \text{ km}}{\text{hr}} \cdot \dfrac{1000 \text{ m}}{1 \text{ km}} \cdot \dfrac{1 \text{ mi}}{1609 \text{ m}}$

$\approx 17,589 \text{ mph}$

49. a.

Length of Side of Box, x	Volume of Cube, x^3	Surface Area, $6x^2$
10	$10^3 = 1000$	$6(10)^2 = 600$
20	$20^3 = 8000$	$6(20)^2 = 2400$
40	$40^3 = 64,000$	$6(40)^2 = 9600$
n	n^3	$6n^2$
2n	$(2n)^3 = 8n^3$	$6(2n)^2 = 6(4)n^2 = 24n^2$

b. The volume becomes 8 times larger.

c. The surface area becomes 4 times larger.

d. $\sqrt{20^2 + 20^2} = \sqrt{2(20^2)} = 20\sqrt{2}$

≈ 28.28